Thermodynamics of Surfaces and Capillary Systems

Chemical Thermodynamics Set

coordinated by
Michel Soustelle

Volume 7

Thermodynamics of Surfaces and Capillary Systems

Michel Soustelle

iSTE

WILEY

First published 2016 in Great Britain and the United States by ISTE Ltd and John Wiley & Sons, Inc.

ISTE Ltd
27-37 St George's Road
London SW19 4EU
UK

www.iste.co.uk

John Wiley & Sons, Inc.
111 River Street
Hoboken, NJ 07030
USA

www.wiley.com

Library of Congress Control Number: 2016939645

British Library Cataloguing-in-Publication Data
A CIP record for this book is available from the British Library
ISBN 978-1-84821-870-3

Contents

Preface

This book – an in-depth examination of chemical thermodynamics – is written for an audience of engineering undergraduates and Masters students in the disciplines of chemistry, physical chemistry, process engineering, materials, etc., and doctoral candidates in those disciplines. It will also be useful for researchers at fundamental- or applied-research labs, dealing with issues in thermodynamics during the course of their work.

These audiences will, during their undergraduate degree, have received a grounding in general thermodynamics and chemical thermodynamics, which all science students are normally taught. This education will undoubtedly have provided them with the fundamental aspects of macroscopic study, but usually the phases discussed will have been fluids exhibiting perfect behavior. Surface effects, the presence of an electrical field, real phases, the microscopic aspect of modeling, and various other aspects, are hardly touched upon (if at all) during this early stage of an academic career in chemical thermodynamics.

This set of books, which is positioned somewhere between an introduction to the subject and a research thesis, offers a detailed examination of chemical thermodynamics that is necessary in the various disciplines relating to chemical or material sciences. It lays the groundwork necessary for students to go and read specialized publications in their different areas. It constitutes a series of reference books that touch on all of the concepts and methods. It discusses both scales of modeling: microscopic (by statistical thermodynamics) and macroscopic, and illustrates the link between them at every step. These models are then used in the study of solid, liquid and gaseous phases, either of pure substances or comprising several components.

The various volumes of the set will deal with the following topics:

– phase modeling tools: application to gases;

– modeling of liquid phases;

– modeling of solid phases;

– chemical equilibrium states;

– phase transformations;

– electrolytes and electrochemical thermodynamics;

– thermodynamics of surfaces, capillary systems and phases of small dimensions.

Appendices in each volume give an introduction to the general methods used in the text, and offer additional mathematical tools and some data.

This series owes a great deal to the feedback, comments and questions from all my students at the *Ecole Nationale Supérieure des Mines* (engineering school) in Saint Etienne who have "endured" my lecturing in thermodynamics for many years. I am very grateful to them, and also thank them for their stimulating attitude. This work is also the fruit of numerous discussions with colleagues who teach thermodynamics in the largest establishments – particularly in the context of the "Thermodic" group, founded by Marc Onillion. My thanks go to all of them for their contributions and kindness.

This seventh instalment is devoted to the study of surface phenomena and to the properties of phases with small dimensions. Chapter 1 looks at the system composed of the interface between a pure liquid and its vapor. A thermodynamic approach is used to determine the influence of the temperature and pressure on the surface tension and its consequences for the specific heat capacities and the latent heats. Chapter 2 describes the modeling and properties of the interfaces between a liquid and a liquid solution or a gaseous mixture. An example of a model of the interface is studied with the model of the strictly-regular solution. Chapter 3 examines the surfaces of solids and solid–solid and solid–liquid interfaces. It closes with the study of electro-capillary phenomena. Chapter 4 deals with small-volume phases, droplets or solids of small dimensions. The thermodynamic values are determined on the basis of Reiss' potential functions. The chapter concludes with a thermodynamic study of the phenomenon of nucleation of a condensed phase. In Chapter 5, we study firstly the thermodynamics of cylindrical capillary, and secondly the properties of thin liquid films. Chapters 6 and 7, respectively, discuss the phenomena of physical adsorption and chemical adsorption of gases by solid surfaces. Finally, in an appendix, we present the application of physical adsorption to the determination of the specific areas of solids and their porosity.

Michel SOUSTELLE
Saint-Vallier
April 2016

Notations and Symbols

A: area of a surface or an interface.

$A_{\mathrm{H}}^{(12)}$: Hamaker constant between two media 1 and 2.

$\mathscr{A}$: affinity.

$\tilde{\mathscr{A}}$: electrochemical affinity.

A_M: molar area.

A_m: molecular area.

a: pressure of cohesion of a gas or radius of the elementary cell of a liquid.

A, B, …: components of a mixture.

b: cosurface of an adsorbed gas.

$\mathscr{E}_p$: set of variables with p intensive variables chosen to define a system.

F: Helmholtz energy.

f_{het}: heterogeneous wetting function.

$\tilde{G}_\sigma$: electrocapillary Gibbs energy.

h_t: spreading coefficient.

h: Planck's constant.

H_{spr}: Harkins spreading coefficient of one liquid over another.

$K_{i,j}(\mathcal{E}_p)$:	thermodynamic coefficient associated with the set of variables $\mathcal{E}p$. X_j is its definition variable and Y_i is its definition function.
K_{ads}:	equilibrium constant of adsorption.
K_{fe}:	equilibrium function of adsorption.
k_B:	Boltzmann's constant.
l_c:	capillary length.
M:	molar mass.
N_a:	Avogadro's number.
N_A:	number of molecules of component A.
P:	pressure of a gas.
p:	spreading parameter.
q_ϕ:	equilibrium heat of adsorption.
q_d:	differential heat of adsorption.
q_{isost}:	isosteric heat of adsorption.
R:	perfect gas constant.
R:	mean radius of curvature of a surface.
r_c:	radius of a cylindrical tube.
r_K:	Kelvin radius.
T:	temperature.
v_{mono}:	volume of a monolayer of adsorbed gas.
$x_k^{(\alpha)}$:	molar fraction of the component k in the α phase.
x_i:	molar fraction of the component i in a solution. Y_i and X_i: conjugate intensive and extensive values.
$y_{i,j}$:	Mayer function.
$\Gamma(\mathcal{E}_p)$:	characteristic function with the set $\mathcal{E}_p$ as canonical variables.
Γ:	characteristic function.

Γ_i:	excess surface or surface concentration of component i.
$\Gamma_{i,j}$:	excess surface or surface concentration of component i in relation to j.
$\gamma_i^{(I)}$:	activity coefficient of component i in the pure-substance reference.
$\gamma_i^{(II)}$:	activity coefficient of component i in the infinitely dilute-solution reference.
$\gamma_i^{(III)}$:	activity coefficient of component i in the molar-solution reference.
$\Delta\sigma$:	spreading on a liquid.
$\Delta_r(A)$:	value of A associated with the transformation r.
θ:	fraction of coverage.
θ_i:	surface fraction of a component.
σ:	surface energy.
σ_e:	surface density of electrical charges.
σ^*:	surface tension.

1

Liquid Surfaces

An interface constitutes an extensive, two-dimensional defect in a system. Given that at least one of the intensive values of that system (as is often the case, for example, with the refractive index) evidently undergoes a discontinuity at that interface, the interface separates two distinct phases. Hence, the system is heterogeneous. The presence of that defect, at least in its vicinity, leads to the modification of the properties of the two phases thus separated. This leads us to model the system considering three phases: two so-called massive (or bulk) phases, which are the phases separated by the interface, and a superficial (surface) phase constituting a layer of a certain thickness, containing the modified properties of the two massive phases. Unlike the two massive phases, which each have their own thermodynamic properties with their own specific thermodynamic coefficients, the surface phase has thermodynamic properties that are dependent on the properties of the two phases surrounding it. Thus, we say that the *surface phase is not autonomous*.

It is common to speak of the surface of a liquid, but in fact this is a misuse of language. In reality, that surface is never isolated from another phase, so in nature we only ever actually find interphases. For example, if the liquid is placed in a vacuum, it vaporizes spontaneously (and least in part), and we see the presence of an interphase between the liquid and its vapor which, in the case of a pure substance, have the same composition but different molecular densities. In this particular case of the equilibrium between a pure substance and its vapor, we sometimes speak of the surface of the liquid, and the properties of that interface are qualified as being the properties of the surface of the liquid. This chapter will be devoted to interfaces between a pure liquid and its vapor.

The different molecular densities of the two bulk phases will lead to anisotropic bond forces in the surface phase. Indeed, the molecules of the liquid which are at the surface are on half of the space in the vicinity of other molecules placed at greater distances, and therefore create an intermolecular force field which also undergoes a discontinuity.

The interface between a pure liquid and its vapor is characterized by easy mechanical deformation and easy variation of its areas. Indeed, we simply need to tilt a recipient to extend the area of the interface separating two fluid phases – i.e. increase the quantity of material making up that interface. This augmentation in the area of the liquid–vapor interface takes place without deformation, because the stresses likely to be engendered are quickly relaxed because the shearing modulus of a liquid is zero.

NOTE.– It is impossible to construct an interface between two pure liquids because reciprocal dissolution, even slight, leads to an interface between two solutions, which will be discussed in Chapter 2.

1.1. Mechanical description of the interface between a liquid and its vapor

Numerous experiments in mechanics show the existence of forces acting on the surface of the liquid in the presence of its vapor. The resultant of those forces seems to be parallel to the surface and tends to reduce the area of the interface.

1.1.1. *Gibbs' and Young's interface models*

To apply mechanics and thermodynamics to interfaces, it is useful to have a model of that interface. The simplest model is Gibbs', whereby the interface is considered to be reduced to the surface of separation of the two phases, with no thickness. In that model, the discontinuity of an intensive value upon the changing phase is sudden, as illustrated by Figure 1.1, which shows the discontinuity of the density on phase change. In order to take account of a certain number of phenomena which we encounter in the study of systems with multiple components, such as adsorption, segregation or surface excess, it is necessary to accept that the surface contains a certain amount of virtual material (a certain number of moles) of each of the species involved.

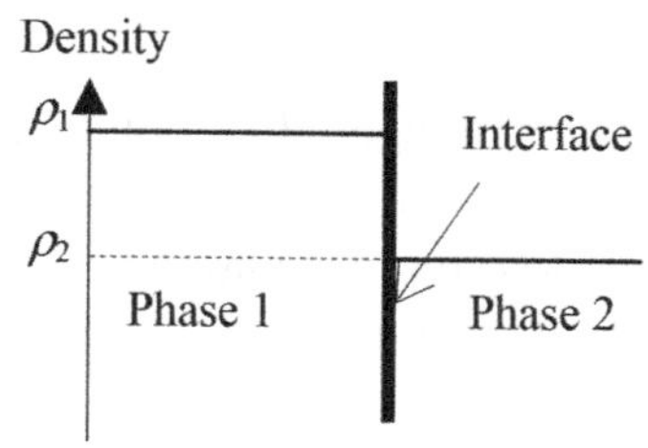

Figure 1.1. *Discontinuity in density in Gibbs' model*

A second, more elaborate, model is Young's layered model. In this model, the interface has a certain thickness or depth, *d*, which is unknown but is likely to be small (see Figure 1.2(a)), at around a few atomic layers, except in the vicinity of the critical point for the liquid–vapor interface.

In Young's model, we cut that surface perpendicularly with a plane AB whose breadth is δl. Figure 1.2(b) illustrates the different forces acting on the left-hand side of the plane AB (with the right-hand side being subject to the same symmetrical forces).

– Between A and A', the force is exerted by the hydrostatic pressure *P''* of the lower phase;

– Between B' and B, the force results from the hydrostatic pressure *P'* in the upper phase;

– Between A' and B', the forces are distributed in accordance with an unknown law.

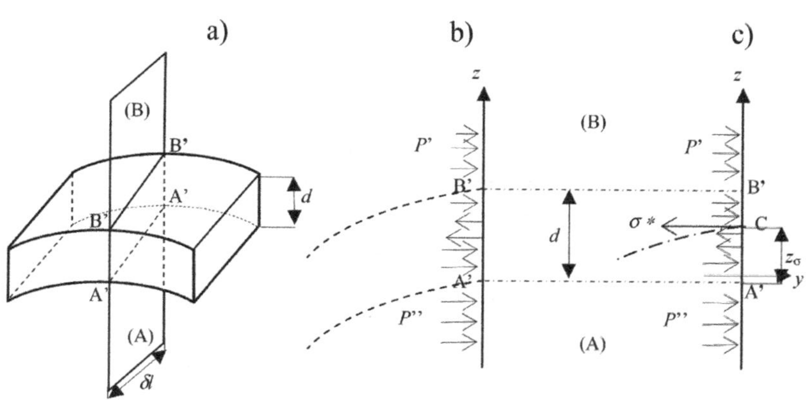

Figure 1.2. *Representation of an interface in Young's model*

Young models the system (see Figure 1.2(c)) as the existence, between B' and A', of a *surface tension* σ^* tangent to a point C, at a distance z_c from A' and such that the equivalences of the forces and the moments in relation to A' are assured between the two representations 1.2*b* and 1.2*c*, which we can express for the forces along the *z* axis by:

$$\sum_z (\vec{F}_z)_b = \sum_z (\vec{F}_z)_c \qquad [1.1]$$

and for the moments in relation to A', by:

$$\sum_z (\vec{M}_{A'})_b = \sum_z (\vec{M}_{A'})_c \qquad [1.2]$$

Between A and C, the forces are due to the pressure *P*', and between C and B they are due to the pressure *P*''.

1.1.2. *Mechanical definition of the surface tension of the liquid*

Let us look again at Young's model for the interface between a pure liquid phase and its vapor. If we extend the free surface of the liquid over a breadth δx (Figure 1.3), the variation in the area of that surface is:

$$\mathrm{d}A = \delta x . \delta l \qquad [1.3]$$

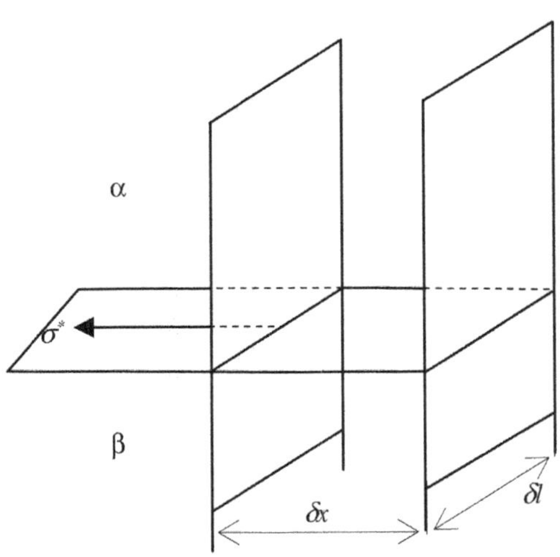

Figure 1.3. *Extension of a portion of surface of a liquid*

The force exerted against the surface tension is:

$$F = \sigma^* \delta l \qquad [1.4]$$

The work which must be injected is the product of that force by the displacement δx. That work will be:

$$\mathrm{d}W = F\delta x = \sigma^* \delta l.\delta x = \sigma^* \mathrm{d}A \qquad [1.5]$$

The term σ^* is called the *surface tension* or *interfacial tension* of the liquid. This value is expressed in Newtons per meter, as shown by relation [1.4].

1.1.3. *Influence of the curvature of a surface – Laplace's law*

Consider an element $\mathrm{d}\vec{S}$ of a curved interface with radii of primary curvatures (in two orthogonal directions) R_1 and R_2 (see Figure 1.4). Each boundary line of that element is subject to forces of surface tension exerted by the rest of the interface.

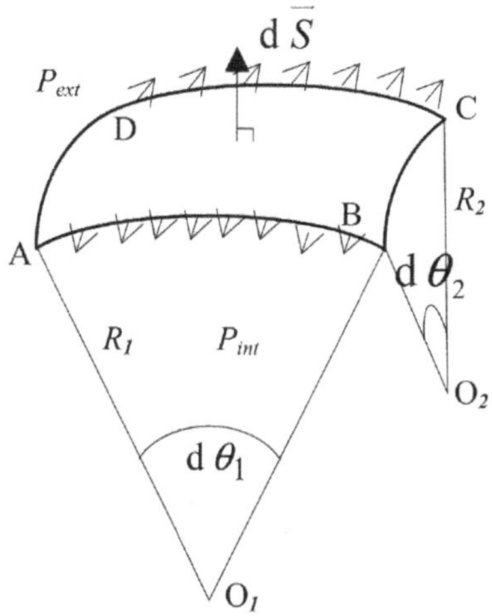

Figure 1.4. *Radii of curvature of a curved surface*

At mechanical equilibrium, the resultant of these forces is canceled out by the forces exerted on the surface by the pressure P_{int} inside the curve and P_{ext} outside of it. As the tangential components, two by two, cancel one another out, it is easy to calculate the normal components. Thus, for instance, on the side AB, the force experienced by the surface element is:

$$-R_1\,\mathrm{d}\theta_1\sigma^*\sin\frac{\mathrm{d}\theta_2}{2}\approx-\frac{1}{2}R_1\,\mathrm{d}\theta_1\sigma^*\,\mathrm{d}\theta_2 \qquad [1.6]$$

The projection of the resultant of all the components, which takes the value of 0, is written:

$$-R_1\,\mathrm{d}\theta_1\,\mathrm{d}\theta_2\sigma^*-R_2\,\mathrm{d}\theta_2\,\mathrm{d}\theta_1\sigma^*+(P_{\text{int}}-P_{\text{ext}})R_1\,\mathrm{d}\theta_1R_2\,\mathrm{d}\theta_2=0 \qquad [1.7]$$

From this, we deduce:

$$P_{\text{int}}-P_{\text{ext}}=\sigma^*\left(\frac{1}{R_1}+\frac{1}{R_2}\right) \qquad [1.8]$$

This is Laplace's law, which gives the expression of the discontinuity in pressure on either side of a curved interface as a function of the surface tension and of the primary radii of curvature of that curved surface.

This law can be expressed in a different form, if we define the mean radius of curvature R by the relation:

$$\frac{1}{R_1}+\frac{1}{R_2}=\frac{2}{R} \qquad [1.9]$$

Laplace's law becomes:

$$P_{\text{int}}-P_{\text{ext}}=\frac{2\sigma^*}{R} \qquad [1.10]$$

Two particular cases of relation [1.8] are often used.

For a spherical surface, such as a drop of liquid, the primary radii of curvature are equal to the radius r of the sphere:

$$R_1=R_2=r \qquad [1.11]$$

and Laplace's law becomes:

$$P_{\text{int}} - P_{\text{ext}} = \frac{2\sigma^*}{r} \qquad [1.12]$$

If we now consider a cylindrical surface with radius r, the primary radii are:

$$R_1 = r \qquad [1.13a]$$

R_2 is infinite [1.13b]

and Laplace's law then takes the form:

$$P_{\text{int}} - P_{\text{ext}} = \frac{\sigma^*}{r} \qquad [1.14]$$

We shall use relations [1.12] and [1.14] in Chapters 4 and 5, which are devoted to the study of phases of small dimensions.

1.2. Thermodynamic approach to the liquid–vapor interface

Considering that the surface work is given by the product of the area by an intensive value σ called the *surface energy*, here we shall discuss a thermodynamic approach to the study of interfaces which, amongst other things, will help us distinguish, in liquids, between the surface tension σ^* as defined by relation [1.5] on the basis of mechanics and the surface energy σ derived from thermodynamics.

1.2.1. *Potential functions*

Let us look again at the layered model shown in Figure 1.2(a), whereby the interface is defined using three volumes: that of the liquid phase, known as the α phase; that of the vapor phase, known as the β phase; and that of the interfacial layer, called the γ phase. The total volume of the system is the sum of those three volumes:

$$V = V^{(\alpha)} + V^{(\beta)} + V^{(\gamma)} \qquad [1.15]$$

The same is true for the other extensive functions, which will all be the sum of three terms – e.g. the internal energy, which would be:

$$U = U^{(\alpha)} + U^{(\beta)} + U^{(\gamma)} \quad [1.16]$$

or the entropy:

$$S = S^{(\alpha)} + S^{(\beta)} + S^{(\gamma)} \quad [1.17]$$

and the quantities of material:

$$n = n^{(\alpha)} + n^{(\beta)} + n^{(\gamma)} \quad [1.18]$$

The extensive variables defining a mole of the system form the set $\mathcal{E}_U$, such that:

$$\mathcal{E}_U = \left\{ S^{(\alpha)} S^{(\beta)}, S^{(\gamma)}, V^{(\alpha)}, V^{(\beta)}, A \right\} \quad [1.19]$$

This set does not contain the volume of the layer $V^{(\gamma)}$, because that latter variable is not independent of the area, A. Indeed, the thickness of the layer is a given property of the substance, and has little dependence on the other variables. By expressing the variation of the internal energy, we obtain:

$$\mathrm{d}U = T\,\mathrm{d}S^{(\alpha)} + T\,\mathrm{d}S^{(\beta)} + T\,\mathrm{d}S^{(\gamma)} - P^{(\alpha)}\,\mathrm{d}V^{(\alpha)} - P^{(\beta)}\,\mathrm{d}V^{(\beta)} + \sigma\,\mathrm{d}A \quad [1.20]$$

The *surface energy* σ, which is the intensive value conjugate to the area, is defined as the partial differential of the internal energy in relation to the area:

$$\sigma = \left(\frac{\partial U}{\partial A} \right)_{S, V^{(\alpha)}, V^{(\beta)}, n} \quad [1.21]$$

The unit in which σ is measured is joules per square meter – i.e. the same dimensions as the surface tension σ^*, which was expressed (see section 1.1.2) in Newtons per meter, which is equivalent to joules per square meter.

If we now choose the set of variables $\mathcal{E}_H$, defined by:

$$\mathcal{E}_H = \left\{ S^{(\alpha)}, S^{(\beta)}, S^{(\gamma)}, P^{(\alpha)}, P^{(\beta)}, A \right\} \qquad [1.22]$$

The potential function would be the enthalpy, defined by:

$$H = U + PV \qquad [1.23]$$

Thus, using relations [1.20] and [1.23], we find the differential of H:

$$\mathrm{d}H = T\,\mathrm{d}S^{(\alpha)} + TS^{(\beta)} + TS^{(\gamma)} + V^{(\alpha)}\,\mathrm{d}P^{(\alpha)} + V^{(\beta)}\,\mathrm{d}P^{(\beta)} + \sigma\,\mathrm{d}A \qquad [1.24]$$

and the surface energy would be such that:

$$\sigma = \left(\frac{\partial H}{\partial A} \right)_{S, P^{(\alpha)}, P^{(\beta)}, n} \qquad [1.25]$$

If we choose the set of variables $\mathcal{E}_F$ defined by:

$$\mathcal{E}_F = \left\{ T, V^{(\alpha)}, V^{(\beta)}, A \right\} \qquad [1.26]$$

the potential function would be the free energy, defined by:

$$F = U - TS \qquad [1.27]$$

Hence, using relations [1.20] and [1.27], we can find the differential of F:

$$\mathrm{d}F = -S^{(\alpha)}\,\mathrm{d}T - S^{(\beta)}\,\mathrm{d}T - S^{(\gamma)}\,\mathrm{d}T - P^{(\alpha)}\,\mathrm{d}V^{(\alpha)} - P^{(\beta)}\,\mathrm{d}V^{(\beta)} + \sigma\,\mathrm{d}A \qquad [1.28]$$

and the surface energy would be such that:

$$\sigma = \left(\frac{\partial F}{\partial A} \right)_{T, V^{(\alpha)}, V^{(\beta)}, n} \qquad [1.29]$$

If we choose the set of variables $\mathcal{E}_G$, defined by:

$$\mathcal{E}_G = \left\{ T, P^{(\alpha)}, P^{(\beta)}, A \right\} \qquad [1.30]$$

the potential function would be the Gibbs energy, defined by:

$$G = U + PV - TS \qquad [1.31]$$

Thus, using relations [1.20] and [1.31], the differential of G would be:

$$\mathrm{d}G = -S^{(\alpha)}\,\mathrm{d}T - S^{(\beta)}\,\mathrm{d}T - S^{(\gamma)}\,\mathrm{d}T + V^{(\alpha)}\,\mathrm{d}P^{(\alpha)} + V^{(\beta)}\,\mathrm{d}P^{(\beta)} + \sigma\,\mathrm{d}A \qquad [1.32]$$

and the surface energy would be such that:

$$\sigma = \left(\frac{\partial G}{\partial A}\right)_{T,P^{(\alpha)},P^{(\beta)},n} \qquad [1.33]$$

If, finally, we choose the set of intensive variables $\mathcal{E}_\sigma$ defined by:

$$\mathcal{E}_\sigma = \left\{T, P^{(\alpha)}, P^{(\beta)}, \sigma\right\} \qquad [1.34]$$

the potential function would be the *capillary Gibbs energy*, defined by:

$$G_\sigma = U + PV - TS - A\sigma = G - A\sigma \qquad [1.35]$$

The differential of G_σ, then, can be defined using relations [1.20] and [1.35]:

$$\mathrm{d}G_\sigma = -S\,\mathrm{d}T + V^{(\alpha)}\,\mathrm{d}P^{(\alpha)} + V^{(\beta)}\,\mathrm{d}P^{(\beta)} - A\,\mathrm{d}\sigma \qquad [1.36]$$

This elementary variation $\mathrm{d}G_\sigma$ corresponds to the elementary work which a transformation is likely to produce, which is deduced from the volume work and the surface work.

Other potential functions can be defined in the same way, by choosing the sets of variables $\left\{T, V^{(\alpha)}, V^{(\beta)}, \sigma\right\}$, $\left\{S^{(\alpha)}, S^{(\beta)}, S^{(\gamma)}, P^{(\alpha)}, P^{(\beta)}, \sigma\right\}$ or $\left\{S^{(\alpha)}, S^{(\beta)}, S^{(\gamma)}, V^{(\alpha)}, V^{(\beta)}, \sigma\right\}$.

1.2.2. *Functions of state of surface*

For the functions relative to the layer, we can define corresponding surface functions of state. For example, for the functions U, H, F and G, we would have the surface values:

$$u^{(\gamma)} = \frac{U^{(\gamma)}}{A} \qquad [1.37a]$$

$$h^{(\gamma)} = \frac{H^{(\gamma)}}{A} \qquad [1.37b]$$

$$f^{(\gamma)} = \frac{F^{(\gamma)}}{A} \qquad [1.37c]$$

$$g^{(\gamma)} = \frac{G^{(\gamma)}}{A} \qquad [1.37d]$$

or indeed, for the surface entropy function:

$$s^{(\gamma)} = \frac{S^{(\gamma)}}{A} \qquad [1.37e]$$

1.2.3. *Equivalence between surface tension and interface energy between two fluids*

We shall now show that, for a liquid, the two values which are expressed in the same dimensions – the surface tension σ^* defined by mechanics and the surface energy σ defined by thermodynamics – are identical.

In order to do this, we consider a closed system with a planar interface between a pure liquid and its vapor. In this case, the pressure is identical in both phases in that volume, which is expressed by:

$$P^{(\alpha)} = P^{(\beta)} = P \qquad [1.38]$$

The amount of material remains constant within the system. Hence, by virtue of relation [1.18], we have:

$$\mathrm{d}n = \mathrm{d}n^{(\alpha)} + \mathrm{d}n^{(\beta)} + \mathrm{d}n^{(\gamma)} = 0 \qquad [1.39]$$

An increase in the area, d*A*, produced by way of a reversible transformation, will require the following isothermal work:

$$\mathrm{d}W = \sigma \, \mathrm{d}A \qquad [1.40]$$

Also, for a transformation taking place at constant temperature, that work is the variation in the free energy in our system, by virtue of relation [1.28], because then:

$$\mathrm{d}F = \sigma^* \, \mathrm{d}A = \mathrm{d}W \qquad [1.41]$$

By comparing relations [1.5] and [1.40], for a liquid phase in the presence of its own gaseous phase, we immediately find the strict equivalence:

$$\sigma^* = \sigma \qquad [1.42]$$

This equivalence between the mechanical and thermodynamic aspects explains why the value σ is indiscriminately called the *surface tension* or the *surface energy* of the liquid in question.

NOTE.– This equivalence is demonstrated only for contact between the fluid phases (independently of their composition), and no longer holds true when the interface is limited by a solid surface undergoing elastic deformation (see Shuttleworth's relation in section 3.1).

1.2.4. *Sign of the energy associated with the surface of a pure liquid*

We can now show how to determine the sign of the surface tension of a liquid in the presence of its own vapor.

The condensation of a vapor into a liquid is always an exothermic phenomenon – ($\Delta_{\text{vap}\to\text{liq}}(H) < 0$) – which is understandable, because the intermolecular bonds are stronger and more numerous (per molecule) in the liquid, which is denser than the vapor.

The bonds between the molecules of a liquid are not covalent, so it is not necessary to take account of a finite nature on the bond in an aggregate of molecules. On the other hand, the increase in density on condensation entails a gap in entropy, which can be approximated as:

$$\Delta_{\text{vap}\to\text{liq}}(S) = \text{R.ln}\left(\frac{\text{molar volume of the liquid}}{\text{molar volume of the gas}}\right) < 0 \qquad [1.43]$$

Thus, the variation of the Gibbs energy:

$$\Delta_{\text{vap}\to\text{liq}}(G) = \Delta_{\text{vap}\to\text{liq}}(H) - T\Delta_{\text{vap}\to\text{liq}}(S)$$

is negative if the temperature is low, but positive if the temperature is higher than the boiling point at the given pressure. The influence of pressure on the Gibbs energy of formation of a compound can be seen through the molar volumes (see the Clausius–Clapeyron relation).

Let us now examine the case of a molecule situated at the edge of a liquid droplet: the bonds are stronger, or the coordinance is stronger, in the liquid than in the surrounding vapor, which has two consequences:

– A force is exerted on the interface, directed toward the inside of the drop. Incidentally, mechanical equilibrium in light of the effect of that surface tension is the reason for the sphericity of the drop.

– The bond energy of the molecule is intermediary between the values, which correspond respectively to the case of a perfect liquid and that of a gas. If we let $\Delta_{\text{vap}\to\text{surf}}(H)$ denote that enthalpy, we therefore have:

$$\Delta_{\text{vap}\to\text{liq}}(H) < \Delta_{\text{vap}\to\text{surf}}(H) < 0 \qquad [1.44]$$

As an initial approximation, we can consider that the density and the degrees of freedom are identical for a surface molecule and a core molecule, so:

$$\Delta_{\text{vap}\to\text{surf}}(S) = \Delta_{\text{vap}\to\text{liq}}(S) \qquad [1.45]$$

The Gibbs energy of condensation at the interface, therefore, is:

$$\begin{aligned}\Delta_{\text{vap}\to\text{surf}}(G) &= \Delta_{\text{vap}\to\text{surf}}(H) - T\Delta_{\text{vap}\to\text{surf}}(S) \\ &= \Delta_{\text{vap}\to\text{surf}}(H) - T\Delta_{\text{vap}\to\text{liq}}(S)\end{aligned} \qquad [1.46]$$

Hence:

$$\Delta_{\text{vap}\to\text{surf}}(G) = \Delta_{\text{vap}\to\text{liq}}(G) + \left[\Delta_{\text{vap}\to\text{surf}}(H) - \Delta_{\text{vap}\to\text{liq}}(H)\right] \qquad [1.47]$$

The Gibbs energy $\Delta_{\text{vap->Dlt}}(G)$ of condensation of the droplets is obtained by adding together the core and interface terms:

$$\Delta_{\text{vap->Dlt}}(G) = \sum_{\text{core}} \Delta_{\text{vap}\to\text{liq}}(G) + \sum_{\text{interface}} \Delta_{\text{vap}\to\text{surf}}(G) \qquad [1.48]$$

so:

$$\Delta_{\text{vap}\to\text{Dlt}}(G) = \Delta_{\text{vap}\to\text{liq}}(G) + x_{\text{int}}\left[\Delta_{\text{vap}\to\text{surf}}(H) - \Delta_{\text{vap}\to\text{liq}}(H)\right] \qquad [1.49]$$

In this expression, $x_{\text{int}} = N^{(\gamma)} / N$ denotes the molar fraction of the molecules at the surface of the droplet.

With the area of the surface A being expressed in m^2 per mole of condensed product, the molar fraction at the interface is: $x_{\text{int}} = A_M / \text{N}_\text{a}.A_m$, where N_a is Avogadro's constant and A_m is the exposed area at the surface per molecule. The value of that term can be estimated on the basis of the molecular radius, or by molecular mechanics in the case of a complex-shaped molecule. By identification with the definition of the surface energy [1.33], we obtain:

$$\sigma = \frac{\Delta_{\text{vap}\to\text{surf}}(H) - \Delta_{\text{vap}\to\text{liq}}(H)}{\text{N}_\text{a} A_m} \qquad [1.50]$$

We saw earlier on that the numerator must be a positive value, so the surface energy is also a positive value. Thus, the surface tension of a liquid is a positive value.

1.2.5. *Extent of the area of the surface of a liquid*

We shall now show that the area occupied by the interface has a minimum value at equilibrium.

In view of the additive properties of the extensive values, the free energy of the system containing two phases, α and β, with an interface, γ, is written as:

$$F = F^{(\alpha)} + F^{(\beta)} + F^{(\gamma)} \qquad [1.51]$$

Because the number of atoms forming the interface is much smaller than the numbers contained in the two volume phases, we can see that any variation in the interfacial area leads to a negligible variation in the quantities of material of the volume phases, and hence in their free energies, so:

$$\left(\frac{\partial F^{(\alpha)}}{\partial A}\right)_{T,V,n} + \left(\frac{\partial F^{(\beta)}}{\partial A}\right)_{T,V,n} \approx 0 \qquad [1.52]$$

Relation [1.29] therefore gives us:

$$\left(\frac{\partial F}{\partial A}\right)_{T,V,n} \approx \left(\frac{\partial F^{(\gamma)}}{\partial A}\right)_{T,V,n} = \sigma \qquad [1.53]$$

NOTE.– The same reasoning can be applied for the other functions U, S, H, G and G_σ.

For a closed system, with constant temperature and volume and a single component, the function $F^{(\gamma)}$ being homogeneous and of first degree in relation to the quantities of material and thus the area of the interface, we would have:

$$F^{(\gamma)} = A\left(\frac{\partial F^{(\gamma)}}{\partial A}\right)_{T,V,n} \qquad [1.54]$$

Thus, in view of relation [1.53]:

$$F^{(\gamma)} = \sigma A \qquad [1.55]$$

Furthermore, the differentiation of the function F, based on relation [1.51], gives us:

$$\mathrm{d}F = \mathrm{d}F^{(\alpha)} + \mathrm{d}F^{(\beta)} + \mathrm{d}F^{(\gamma)} = \mathrm{d}F^{(\alpha)} + \mathrm{d}F^{(\beta)} + \mathrm{d}(\sigma A) \qquad [1.56]$$

However, at constant temperature, volume and quantity of material, the free energies of the volume phases are constant, and thus in light of relation [1.55]:

$$(\mathrm{d}F)_{T,V,n} = \mathrm{d}(\sigma A)_{T,V,n} \qquad [1.57]$$

As the free energy function is a potential function within the set of variables chosen, at equilibrium we have:

$$\mathrm{d}(\sigma A)_{T,V,n} = 0 \qquad [1.58]$$

For a pure substance, the surface energy is uniform and independent of the area, so the equilibrium condition is simplified to give us:

$$\sigma \mathrm{d}(A)_{T,V,n} = 0 \qquad [1.59]$$

Thus, the area occupied by the interface has its minimum value at equilibrium.

1.3. Influence of temperature on surface energy

Because of the symmetry of the characteristic matrix of the system, it is easy to demonstrate the following relation between the derivative of the surface energy in relation to temperature and the entropy expressed per unit area:

$$\frac{\partial \sigma}{\partial T} = -s_{\sigma} \qquad [1.60]$$

However, for want of a more accurate model, initially the relation between the surface tension and the temperature was based solely on experimental observations and similarities.

Experience tells us that usually the surface energy decreases with increasing temperature, reaching the value of 0 at the critical temperature T_c.

This observation led Eötvös to suggest a simple linear decrease, written as follows as a function of the temperatures, expressed in degrees Celsius:

$$\sigma = \sigma_0 \left(1 - \frac{T}{T_c} \right) \quad [1.61]$$

However, it is clear that this linear law is very soon proven false by experimentation, as we approach the critical temperature, and therefore, van der Waals and Guggenheim were led to correct the previous formula in the form:

$$\sigma = \sigma_0 \left(1 - \frac{T}{T_c} \right)^n \quad [1.62]$$

Eötvös, trying an analogy between the area and the volume and basing the reasoning on the perfect gas model, proposed the relation:

$$\sigma \left(v^0 \right)^{2/3} = k \left(T_c - T \right) \quad [1.63]$$

In view of the molar volume of the gas $v^{0(vap)}$, which is no longer negligible in relation to that of the liquid $v^{0(liq)}$ in the vicinity of the critical point, the previous formula is corrected, and the new proposition from Eötvös and Katayama is written:

$$\sigma \left(\frac{1}{v^{0(liq)}} - \frac{1}{v^{0(vap)}} \right)^{-2/3} = kT_c \left(1 - \frac{T}{T_c} \right) \quad [1.64]$$

By combining relations [1.64] and [1.62] with $n = 1.2$, we obtain a new proposition, constituted by the MacLeod relation:

$$\frac{\sigma^{1/4}}{\left(v^{0(liq)} - v^{0(vap)} \right)^4} - 1 = \frac{k^{3/2} T_c^{3/2}}{\sigma_0^{5/4}} = \left[P \right] \cong \sigma^{1/4} v^{0(liq)} \quad [1.65]$$

Fowler gave a certain statistical basis to this relation. However, in order to progress further, we shall describe the model developed by Metzger, which gives an expression of the surface energy with two adjustable parameters, and which seems satisfactory both for simple liquids and for associated liquids and metals.

On the basis of the definition of the free energy function F and of its derivative in relation to temperature (-S), it is easy to demonstrate the Gibbs–Helmholtz relation:

$$T\frac{\partial F}{\partial T} - F = -U \qquad [1.66]$$

Let us express the internal energy in the form of a limited expansion of the temperature:

$$U = U_0 + \alpha T + \beta T^2 + \gamma T^3 \qquad [1.67]$$

By integration and treating $\frac{\partial F}{\partial T}$ and $\frac{\mathrm{d}F}{\mathrm{d}T}$ as identical, we find:

$$F = U_0 - \alpha T \ln T - \beta T^2 - \frac{\gamma}{2}T^3 + aT \qquad [1.68]$$

In view of relation [1.67] and by derivation, we obtain:

$$\frac{\partial F}{\partial T} = -\alpha - 2\beta T - \frac{3}{2}\gamma T^2 + a - \alpha \ln T \qquad [1.69]$$

The value $\frac{\partial F}{\partial T}$, excepting the sign, is the same as the entropy. Let us bring a molecule of the liquid from the depths of the volume to the surface and apply the above relations to that transformation. We accept that at the temperature of absolute zero, the molecular order is the same at the surface and inside the liquid, so the variation in entropy will be 0 at absolute zero. This being the case, it is easy to see that this means that $\alpha = 0$ and $a = 0$. Relation [1.67] can therefore be simplified to:

$$U = U_0 + \beta T^2 + \gamma T^3 \qquad [1.70]$$

and relation [1.68] is also simplified, giving us:

$$F = U_0 - \beta T^2 - \frac{\gamma}{2} T^3 \qquad [1.71]$$

For the two functions $\Delta_{\text{liq}\to\text{surf}} U$ and $\Delta_{\text{liq}\to\text{surf}} F$, for the next stage of our discussion, we shall limit ourselves to the first two terms in the expansion, meaning that we adopt the two expressions:

$$\Delta_{\text{liq}\to\text{surf}} U = \Delta_{\text{liq}\to\text{surf}} U_0 + \beta T^2 \qquad [1.72a]$$

$$\Delta_{\text{liq}\to\text{surf}} F = \Delta_{\text{liq}\to\text{surf}} U_0 - \beta T^2 \qquad [1.72b]$$

In the liquid, the mean of the intermolecular distances is denoted by r, such that, if $v^{0(liq)}$ denotes the molar molecular volume, we have:

$$v^{0(liq)} = \mathrm{N_a} r^3 \qquad [1.73]$$

In addition, the surface energy is also the amount of work that needs to be injected to increase the free surface by 1 m^2, and has the value σ, as we saw in section 1.1.2.3.

In order to apply relations [1.72a] and [1.72b] to the free surface of the liquid, let us bring a molecule from inside the body of liquid to the surface. The area of the surface increases approximately by a quantity equal to r^2, and the work to be injected is σr^2. That work is also equal to $\Delta_{\text{liq}\to\text{surf}} F = \Delta_{\text{liq}\to\text{surf}} U_0 - \beta T^2$. Hence, in light of relation [1.73], we have the triple equality:

$$\Delta_{\text{liq}\to\text{surf}} F = \Delta_{\text{liq}\to\text{surf}} U - \beta T^2 = \sigma r^2 = \sigma \frac{\left(v^{0(liq)}\right)^{2/3}}{\mathrm{N_a^{2/3}}} \qquad [1.74]$$

From this, we deduce the expression of the surface tension of a liquid or its surface energy:

$$\sigma = \frac{\left(\Delta_{\text{liq}\to\text{surf}} U_0 - \beta T^2\right)\mathrm{N_a^{2/3}}}{\left(v^{0(liq)}\right)^{2/3}} \qquad [1.75]$$

In fact, we have supposed that the molecules of the liquid were compact spheres, and it is simpler to write that the area they occupy is $(r-l)^2$, where l is a correction factor which we need to determine. In these conditions, relation [1.75] is replaced by:

$$\sigma=\frac{\left(\Delta_{\text{liq}\to\text{surf}}U_0-\beta T^2\right)\mathrm{N}_{\mathrm{a}}^{2/3}}{\left(\left(v^{0(liq)}\right)^{1/3}-l\right)^2} \qquad [1.76]$$

We can reduce the number of constants in relation [1.76] by first writing that the surface energy must take the value of 0 at the critical temperature, which is expressed by:

$$\beta=\frac{\Delta_{\text{liq}\to\text{surf}}U_0}{T_c^2} \qquad [1.77]$$

The expression of the surface energy, then, is:

$$\sigma=\frac{\Delta_{\text{liq}\to\text{surf}}U_0\mathrm{N}_{\mathrm{a}}^{2/3}\left(1-\left(\frac{T}{T_c}\right)^2\right)}{\left(\left(v^{0(liq)}\right)^{1/3}-l\right)^2} \qquad [1.78]$$

To determine l and $\Delta_{\text{liq}\to\text{surf}}U_0$, we simply need to apply relation [1.78] for the same substance at two different temperatures on the basis of two experimental surface energy values, so we shall have two equations which allow us to determine the two unknowns $\Delta_{\text{liq}\to\text{surf}}U_0$ and l.

In fact, relation [1.78] overlooks the presence of vapor on top of the liquid. If vaporous molecules are present, they too will exert forces on the surface molecules, which are expressed by a corrective term taking account of the work of those forces, which obey relation [1.78]. The consequence of this is that the effective work is the difference between two terms, which gives us the relation:

$$\sigma = \Delta_{\text{liq}\to\text{surf}} U_0 \mathrm{N}_a^{2/3} \left(1-\left(\frac{T}{T_c}\right)^2\right)\left[\frac{1}{\left(\left(v^{0(liq)}\right)^{1/3}-l\right)^2}-\frac{1}{\left(\left(v^{0(vap)}\right)^{1/3}-l\right)^2}\right] \quad [1.79]$$

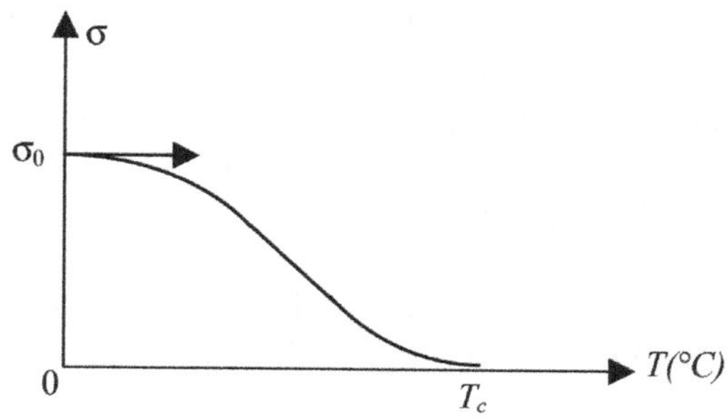

Figure 1.5. *Variations in the surface energy with changing temperature in Metzger's model*

Benzene			Water			Mercury		
T(°C)	σ measured (j/m²)	σ calculated (j/m²)	T(°C)	σ measured (j/m²)	σ calculated (j/m²)	T(°C)	σ measured (j/m²)	σ calculated (j/m²)
0	0.0317	0.0317	0	0.0756	0.0756	30	0.4716	0.471
20	0.02902	0.02905	10	0.0742	0.0744	40	0.4682	0.4669
50	0.0251	0.0251	20	0.0727	0.0730	100	0.4562	0.4541
70	0.0225	0.0226	40	0.0696	0.0700	200	0.4312	0.4306
100	0.0188	0.0189	60	0.0662	0.0665	300	0.3995	0.4051
150	0.0129	0.0129	80	0.0626	0.0627	360	0.3764	0.3887
200	0.0074	0.0075	100	0.0589	0.0590	1450	0	0
250	0.0027	0.0027	130	0.0528	0.0529			
288.5	0	0						

Table 1.1. *Variations in surface tensions of a number of substances with temperature, according to Metzger*

Often, the second term, due to the vapor, is negligible in comparison to the first.

Figure 1.5 plots the variations in the surface tension with changing temperature. Table 1.1 gives a number of values provided by Metzger, which can be used to draw a comparison between the calculated values and the experimental values, in the cases of benzene (which is a simple liquid), of water (which is an associated liquid, in view of the hydrogen bonds within it) and for a metal: mercury. We can see the fairly close correspondence between the measured and calculated values.

Remember that Metzger's fundamental hypothesis states that at absolute zero, there is the same degree of order in the liquid phase and in the surface, meaning there is no difference in entropy at absolute zero, which is reminiscent of Planck's third law.

1.4. Surface latent heat

Consider a transformation at constant pressure, varying only the area of the interface and the temperature. We shall calculate the total surface energy defined by $h_\sigma = \left(\frac{\partial H}{\partial A}\right)_{T,P,n}$. We can write:

$$g_\sigma = h_\sigma - Ts_\sigma \qquad [1.80]$$

Using relations [1.33] and [1.60], we obtain:

$$h_\sigma = \sigma - T\left(\frac{\partial \sigma}{\partial T}\right)_{P,n,A} \qquad [1.81]$$

This equation was given by Kelvin. As σ is positive (see section 1. 2.4) and σ decreases as the temperature increases (see the curve in Figure 1.5), h_σ is positive, meaning that heat is indeed absorbed by a liquid when its surface area increases.

1.5. Surface specific heat capacity

By deriving the relation between the molar specific heat capacity and the enthalpy in relation to the area A, we find:

$$\frac{1}{T}\left(\frac{\partial C_P}{\partial A}\right)_T = \left(\frac{\partial\left(\frac{\partial S}{\partial T}\right)_A}{\partial A}\right)_T = \left(\frac{\partial\left(\frac{\partial S}{\partial A}\right)_T}{\partial T}\right)_A = \left(\frac{\partial s_\sigma}{\partial T}\right)_A \qquad [1.82]$$

Taking account of relation [1.60], we obtain:

$$\left(\frac{\partial C_P}{\partial A}\right)_T = -T\left(\frac{\partial^2 \sigma}{\partial T^2}\right)_A \qquad [1.83]$$

This equation connects the variation in the specific heat capacity at constant pressure with the area of the interface to the second derivative of the surface energy in relation to temperature.

If we examine the curve in Figure 1.5, we can roughly distinguish three zones:

– The first zone, in which, at low temperatures, the second derivative of the surface energy in relation to the temperature is negative. In that zone, by virtue of relation [1.83], the specific heat capacity at constant pressure increases with the area of the interface, and more heat is needed to raise the temperature of the surface molecules by 1° than to produce the same temperature in the molecules within the liquid.

– The second zone, where, at a medium temperature, the second derivative of the surface energy in relation to temperature has the value of 0. In that zone, by virtue of relation [1.83], the specific heat capacity at constant pressure is independent of the area of the interface, so the same amount of heat is needed to raise the temperature by 1° both for the molecules at the surface and for those within the liquid.

– The third zone, in which, at a high temperature, the second derivative of the surface energy in relation to the temperature is positive. In that zone, by virtue of relation [1.83], the specific heat capacity at constant pressure decreases with the area of the interface. Hence, less heat is needed to raise the temperature of the surface molecules by 1° than for the molecules within the liquid.

However, we generally see that the surface term of the specific heat capacity is lower than the degree of accuracy of the measurements, which means that often this contribution is not taken into account.

1.6. Influence of pressure on the surface tension of a liquid

It is only possible to increase the pressure above a liquid interface with its vapor if the gaseous phase contains a foreign gas which is insoluble in the liquid. In such a case, the surface energy varies with the pressure of that gas. Experience shows us that the surface energy of the liquid is not altered for low pressures of around 1 atmosphere, but that this variation is far from negligible if we use higher pressures. Kindt showed that this variation can be as great as 50% if we use pressure values of 150 atmospheres.

Due to the symmetry of the characteristic matrix, we have:

$$\frac{\partial \sigma}{\partial P} = -\frac{\partial V}{\partial A} \qquad [1.84]$$

Thus, the variation of the surface energy with pressure is, in fact, a form factor of the liquid.

In his model, Metzger (see section 1.3) included the pressure variable as follows.

If we use relation [1.72], when a molecule comes from the interior of the liquid to the surface, the free energy is increased by $P\Delta v$, where Δv is the variation in the volume of the molecule transferred, so that the free energy becomes:

$$\Delta_{\text{liq}\rightarrow\text{surf}} F = \Delta_{\text{liq}\rightarrow\text{surf}} U_0 - \beta T^2 + P\Delta v \qquad [1.85]$$

Obeying the same process of reasoning as in section 1.3, relation [1.79] is substituted by:

$$\sigma = \left[\Delta_{\text{liq}\to\text{surf}} U_0 \left(1 - \left(\frac{T}{T_c} \right)^2 \right) + P\Delta v \right] \left[\frac{1}{\left(\left(v^{0(liq)} \right)^{1/3} - l \right)^2} - \frac{1}{\left(\left(v^{0(vap)} \right)^{1/3} - l \right)^2} \right] \mathrm{N}_a^{2/3} \quad [1.86]$$

This expression accounts for the fact that if the pressure is low, the inert gas has no influence on the surface energy. At high pressure, on the other hand, its influence becomes notable.

1.7. Evaluation of the surface energy of a pure liquid

To evaluate the surface energy of a liquid, we need to have a microscopic model at our disposal. The simplest available microscopic model is that of the monomolecular surface layer.

In this model, we consider that the surface of the liquid is formed of a monomolecular layer covering the volume of the liquid.

The model is based on the following hypotheses:

– the potential energy between two near-neighboring molecules $\varepsilon_{AA}(r)$ (negative) is a rapidly-decreasing function of r, which means we can ignore the interactions between molecules that are not near neighbors;

– the molecules of the liquid are arranged in a quasi-net, so for each molecule, we can define a mean value z for the number of near neighbors.

Within the liquid volume phase, imagine there are molecular levels parallel to the surface, represented by layers 1, 2 and 3 in Figure 1.6. A molecule A in layer 2, for instance, has near neighbors which are either in layer 2 itself or in the two adjacent layers, 1 and 3. Reasoning on the basis of the molar values, each layer contains a mole – i.e. N_a molecules (N_a is Avogadro's number).

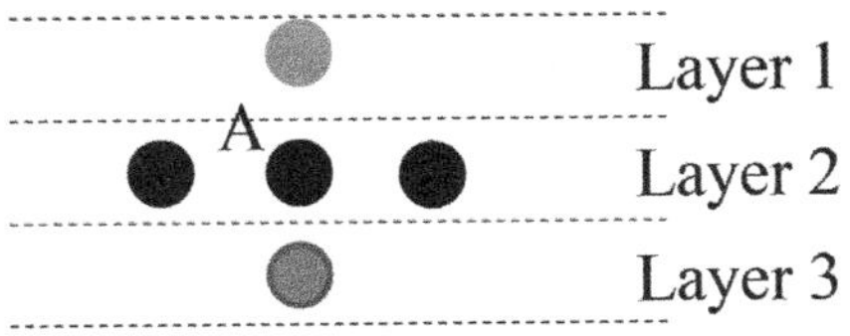

Figure 1.6. *Layers within a liquid, parallel to the surface*

Let lz denote the number of molecules that are the nearest neighbors of a molecule and situated in the same layer as that molecule, whilst mz represents the number of near neighbors situated in a layer adjacent to that containing the molecule. We have:

$$lz + 2mz = z \qquad [1.87a]$$

so:

$$l + 2m = 1 \qquad [1.87b]$$

For instance, if we choose the cubic stack with centered faces, we have $z = 12$, $m = 1/4$ and $l = 1/2$.

If we imagine that the liquid is split parallel to the layers defined previously, then we create two free surfaces with the areas A_M (molar area), each containing Na molecules. The excess energy caused by that split would be twice the surface energy, and thus:

$$\sigma = \frac{zm\mathrm{N_a}\varepsilon_{AA}}{2A_M} \qquad [1.88]$$

This energy can be linked to the enthalpy of vaporization of the liquid, because vaporization breaks all the intermolecular bonds. Thus, we have:

$$\varepsilon_{AA} = \frac{2\Delta_{\mathrm{liq}\to\mathrm{vap}}H}{z\mathrm{N_a}} \qquad [1.89]$$

and hence the surface energy is:

$$\sigma = \frac{m\Delta_{\mathrm{liq}\to\mathrm{vap}}H}{A_M} \qquad [1.90]$$

In fact, the split has created interfaces between the liquid and the vacuum, whereas the surface tension is defined in the presence of the vapor of the liquid. The calculation presented above, though, is correct because the distances between the gas molecules and the monomolecular surface layer are sufficiently large for us to ignore the interaction energies.

We note that the surface energies calculated by relations [1.88] or [1.90] are overestimated. One of the main criticisms leveled at the model is that it is accepted that the surface layer has the same structure as a monomolecular layer situated within the liquid.

In order to bring a molecule from inside the liquid to the surface, it is necessary to break the *mz* bonds, requiring a so-called energy of extension. We can define the *Stéphan ratio* as the ratio of the enthalpy of vaporization to the molar energy of extension. In our model, that ratio is *z/mz* – i.e. *1/m*. This means, for example, that for a centered-face cubic net, that ratio would be 4 – a value which is much too high – so for example, for liquid argon, that ratio is only 2.38. This means that, unlike with our hypothesis and that employed by Metzger (see section 1.3), the surface phase would not have the same degree of order, and therefore the same entropy, as a liquid monolayer. The surface layer would be much more disordered than a monolayer of the liquid phase, and in particular, it would contain a great many vacancies.

2

Interfaces Between Liquids and Fluid Solutions

In Chapter 1, we limited our discussion to liquid–fluid interfaces with a single component – in other words, interfaces between a pure liquid and its vapor. We shall now examine the properties of interfaces between a liquid and a fluid (liquid or gas) containing multiple components. The components will be denoted by the indices 1, 2, … *i*, …. In this case, the layer modeling the interface is the site of phenomena of adsorption, which depend on the composition of each of the phases in the bulk, and thus unlike those bulk phases whose properties are independent of their neighbors, the properties of the layer depend on the variables which determine the properties of the neighboring bulk layers. We say that the layer is a *non-autonomous phase*, whilst the bulk phases on both sides of the layer are *autonomous phases*.

2.1. Surface concentrations and surface excess

In the layer, the position of the interface is theoretically determined by the difference in pressure, $P^{(\alpha)} - P^{(\beta)}$, which sets the mean radius of curvature of the surface. The layer contains a certain amount $n_i^{(\gamma)}$ of each component M_i of the system. The surface concentration Γ_i of the species M_i, also called the *absolute adsorption* or *surface excess*, will therefore be:

$$\Gamma_i = \frac{n_i^{(\gamma)}}{A} \qquad [2.1]$$

We shall now evaluate $n_i^{(\gamma)}$ or Γ_i. In order to do this, consider two phases in contact, α and β, with the respective volumes $V^{(\alpha)}$ and $V^{(\beta)}$, separated by the interface Σ (Figure 2.1). Let $C_i^{(\alpha)}$ and $C_i^{(\beta)}$ be the molar concentrations of component M_i in each of the bulk phases. If those phases remain uniform up to the separating surface Σ, they will contain $n_i^{(\alpha)} = C_i^{(\alpha)}V^{(\alpha)}$ and $n_i^{(\beta)} = C_i^{(\beta)}V^{(\beta)}$ moles of M_i respectively. If n_i is the quantity of M_i contained in the whole system, by the material balance we can write:

$$n_i^{(\gamma)} = n_i - \left(n_i^{(\alpha)} + n_i^{(\beta)}\right) \qquad [2.2]$$

This expression explains the term *total surface excess relative to component M_i*. That excess may be positive, in which case we have *positive adsorption* of M_i, or negative, which leads to *negative adsorption*.

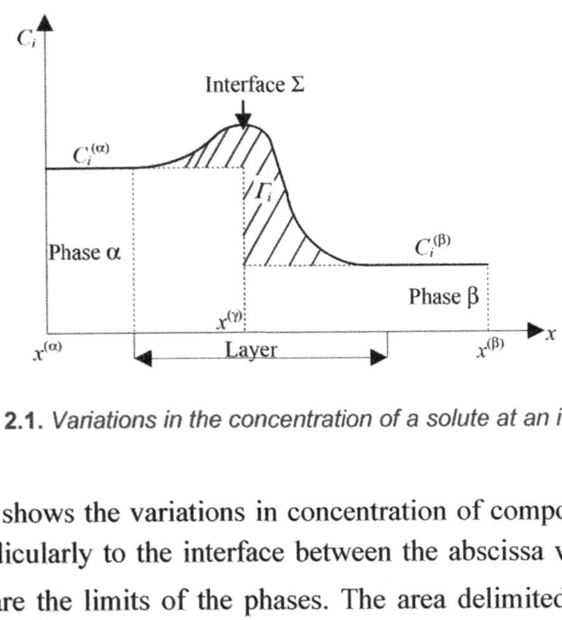

Figure 2.1. *Variations in the concentration of a solute at an interface*

Figure 2.1 shows the variations in concentration of component M_i as we move perpendicularly to the interface between the abscissa values $x^{(\alpha)}$ and $x^{(\beta)}$, which are the limits of the phases. The area delimited by that curve represents the total quantity of component M_i contained in a cylinder with surface of 1 and height $x^{(\beta)} - x^{(\alpha)}$. Γ_i is the area of the hatched surface. The figure is constructed in the case of positive adsorption.

The layered model, though, presents a problem in terms of the choice of abscissa $x^{(\gamma)}$ of the interface. We have defined it in relation to the discontinuity of the pressure, which sets the radii of curvature. However, this determination is not very accurate, and even less so when the radii of curvature are high. The position is no longer defined in the case of planar interfaces, for which there is no longer a pressure discontinuity. However, we can see in Figure 2.1 that the value of Γ_i depends heavily on the position of the interface, and can even turn a positive value into a negative one simply by changing the value of $x^{(\gamma)}$. For this reason, we are led to define, instead of the absolute adsorption Γ_i, a *relative adsorption* $\Gamma_{i,1}$ of component M_i in relation to component M_1, for which it is helpful to choose the solvent, if it exists.

The quantity of material $n_i^{(\gamma)}$ of component M_i in the layer can be written in the form:

$$n_i^{(\gamma)} = n_i - \left(C_i^{(\alpha)} V^{(\alpha)} + C_i^{(\beta)} V^{(\beta)}\right) \tag{2.3}$$

By introducing the total volume of the system ($V = V^{(\alpha)} + V^{(\beta)}$), relation [2.3] becomes:

$$n_i^{(\gamma)} = n_i - C_i^{(\alpha)} V + \left(C_i^{(\alpha)} - C_i^{(\beta)}\right) V^{(\beta)} \tag{2.4}$$

The volume V is independent of the choice of position of the interface, but volume $V^{(\beta)}$ does depend upon it. If we eliminate that volume between the two relations [2.4] written for component M_i, on the one hand, and component M_1 on the other, we obtain:

$$\frac{n_i^{(\gamma)} - \left(n_i - C_i^{(\alpha)} V\right)}{C_i^{(\alpha)} - C_i^{(\beta)}} = \frac{n_1^{(\gamma)} - \left(n_1 - C_1^{(\alpha)} V\right)}{C_1^{(\alpha)} - C_1^{(\beta)}} \tag{2.5}$$

or

$$n_i^{(\gamma)} - n_1^{(\gamma)} \frac{C_i^{(\alpha)} - C_i^{(\beta)}}{C_1^{(\alpha)} - C_1^{(\beta)}} = \left(n_i - C_i^{(\alpha)} V\right) - \left(n_1 - C_1^{(\alpha)} V\right) \frac{C_i^{(\alpha)} - C_i^{(\beta)}}{C_1^{(\alpha)} - C_1^{(\beta)}} \tag{2.6}$$

As the right-hand side of relation [2.6] is independent of the position of the interface, the same is true of the left-hand side. If we divide by the area A, we define the relative adsorption $\Gamma_{i,1}$ of component M_i in relation to component M_1 by the relation:

$$\Gamma_{i,1} = \Gamma_i - \Gamma_1 \frac{C_i^{(\alpha)} - C_i^{(\beta)}}{C_1^{(\alpha)} - C_1^{(\beta)}} \qquad [2.7]$$

As we can see, the relative adsorption does not depend on the position of the interface.

We can see from the defining relation [2.7] that if $\Gamma_1 = 0$, we have $\Gamma_{i,1} = \Gamma_i$, so the relative adsorption of component M_i in relation to component M_1 is the absolute adsorption of M_i, when the position of the interface is chosen such that the adsorption of M_1 is null.

We can also see that if the reference component is the solvent and if the solutions are sufficiently dilute, meaning that if we simultaneously have $C_i^{(\alpha)} << C_1^{(\alpha)}$ and $C_i^{(\beta)} << C_1^{(\beta)}$, the relative adsorption $\Gamma_{i,1}$ is practically identical to the absolute adsorption Γ_i.

NOTE.– From relation [2.7], we can deduce that the relative adsorption $\Gamma_{1,1}$ of a component in relation to itself is zero.

A particular case of relation [2.7] is obtained if one of the phases – say, the β phase, for example – is a gas. We can then overlook the concentrations in that phase in comparison to those in the liquid phase, expressed by $C_i^{(\beta)} << C_i^{(\alpha)}$ and $C_1^{(\beta)} << C_1^{(\alpha)}$, and relation [2.7] becomes:

$$\Gamma_{i,1} = \Gamma_i - \Gamma_1 \frac{C_i^{(\alpha)}}{C_1^{(\alpha)}} = \Gamma_i - \Gamma_1 \frac{x_i^{(\alpha)}}{x_1^{(\alpha)}} \qquad [2.8]$$

$x_i^{(\alpha)}$ is the molar fraction of component i in the liquid α phase. If, furthermore, component 1 is the solvent in a dilute solution, then a new simplification leads to a relative adsorption which varies in a linear fashion with the molar fraction or the concentration:

$$\Gamma_{i,1} \cong \Gamma_i - \Gamma_1 C_i^{(\alpha)} v_1^{0(\alpha)} \cong \Gamma_i - \Gamma_1 x_i^{(\alpha)} \qquad [2.9]$$

As formula [2.8] is valid irrespective of the position of the separating surface, let us place it at the lower limit of the layer – i.e. at the boundary between the layer and the α phase. The concentrations Γ_i and Γ_1 are equal to the quantities of i and 1 per unit surface in the layer. This means that the relative adsorption of i in relation to 1 is null when the components i and 1 are present in the layer in the same proportions as in the solution. Thus, the relative adsorption expressed the fact that the interface is enriched or impoverished in terms of the various components in relation to the core of the solution. This is consistent with the fact that the relative adsorption of a component in relation to itself is zero.

2.2. Thermodynamics of interfaces of polycomponent liquid–fluid systems

As we did for pure substances (section 1.2), we shall now discuss the thermodynamics of surfaces for polycomponent systems.

2.2.1. *Complete chemical potential of a component in a phase*

In the context of the layered model (section 1.2.1), the first set of variables, which are all extensive, will be:

$$\mathcal{E}_U = \left\{ S, V^{(\alpha)}, V^{(\beta)}, A, n_i^{(\alpha)}, n_i^{(\beta)}, n_i^{(\gamma)} \right\} \qquad [2.10]$$

In this set of variables, the differential of the internal energy is written:

$$\begin{aligned} \mathrm{d}U &= T\,\mathrm{d}S - P^{(\alpha)}\,\mathrm{d}V^{(\alpha)} - P^{(\beta)}\,\mathrm{d}V^{(\beta)} + \sigma\,\mathrm{d}A \\ &\quad + \sum_i \frac{\partial U}{\partial n_i^{(\alpha)}}\mathrm{d}n_i^{(\alpha)} + \sum_i \frac{\partial U}{\partial n_i^{(\beta)}}\mathrm{d}n_i^{(\beta)} + \sum_i \frac{\partial U}{\partial n_i^{(\gamma)}}\mathrm{d}n_i^{(\gamma)} \end{aligned} \qquad [2.11]$$

This differential includes the partial differentials $\dfrac{\partial U}{\partial n_i^{(\alpha)}}$, $\dfrac{\partial U}{\partial n_i^{(\beta)}}$ and $\dfrac{\partial U}{\partial n_i^{(\gamma)}}$. These differentials, which resemble chemical potentials, are known as the *complete chemical potentials* or *surface chemical potentials* of component i, in the α, β and γ phases respectively. Thus, by definition of those complete chemical potentials, we have:

$$\left[\mu_i^{(\alpha)}\right]=\left(\frac{\partial U}{\partial n_i^{(\alpha)}}\right)_{S,V^{(\alpha)},V^{(\beta)},A,n_j} \qquad [2.12]$$

and similar relations are written for the other phases. Thus, now the differential of the internal energy will be:

$$\begin{aligned}\mathrm{d}U &= T\,\mathrm{d}S - P^{(\alpha)}\,\mathrm{d}V^{(\alpha)} - P^{(\beta)}\,\mathrm{d}V^{(\beta)} + \sigma\,\mathrm{d}A \\ &+\sum_i\left[\mu_i^{(\alpha)}\right]\mathrm{d}n_i^{(\alpha)}+\sum_i\left[\mu_i^{(\beta)}\right]\mathrm{d}n_i^{(\beta)}+\sum_i\left[\mu_i^{(\gamma)}\right]\mathrm{d}n_i^{(\gamma)}\end{aligned} \qquad [2.13]$$

The same reasoning can be applied for the functions H, F, G, and G_σ, each in their set of variables – e.g. for the differential of the Helmholtz energy, we would have:

$$\begin{aligned}\mathrm{d}F &= -S\,\mathrm{d}T - P^{(\alpha)}\,\mathrm{d}V^{(\alpha)} - P^{(\beta)}\,\mathrm{d}V^{(\beta)} + \sigma\,\mathrm{d}A \\ &+\sum_i\left[\mu_i^{(\alpha)}\right]\mathrm{d}n_i^{(\alpha)}+\sum_i\left[\mu_i^{(\beta)}\right]\mathrm{d}n_i^{(\beta)}+\sum_i\left[\mu_i^{(\gamma)}\right]\mathrm{d}n_i^{(\gamma)}\end{aligned} \qquad [2.14]$$

and thus, we can write the expression of a complete chemical potential in the following different forms:

$$\begin{aligned}\left[\mu_i^{(\alpha)}\right] &= \left(\frac{\partial U}{\partial n_i^{(\alpha)}}\right)_{S,V^{(\alpha)},V^{(\beta)},A,n_j} = \left(\frac{\partial H}{\partial n_i^{(\alpha)}}\right)_{S,P^{(\alpha)},P^{(\beta)},A,n_j} \\ &= \left(\frac{\partial F}{\partial n_i^{(\alpha)}}\right)_{T,V^{(\alpha)},V^{(\beta)},A,n_j} = \left(\frac{\partial G}{\partial n_i^{(\alpha)}}\right)_{T,P^{(\alpha)},P^{(\beta)},A,n_j} = \left(\frac{\partial G_\sigma}{\partial n_i^{(\alpha)}}\right)_{T,P^{(\alpha)},P^{(\beta)},\sigma,n_j}\end{aligned} \qquad [2.15]$$

As is demonstrated by this last equation, the complete chemical potential is the generalized chemical potential for surface phenomena.

Relations [1.21], [1.25], [1.29] and [1.33], defining the surface energy, remain valid, with the partial differentials being taken at constant quantities of the species in the different phases.

NOTE.– As the functions H, F, G, and G_σ are homogeneous first-degree functions in relation to the quantities of material, the application of Euler's theorem gives us:

$$U = TS - PV + \sigma A + \sum_i \left[\mu_i^{(\alpha)}\right] n_i^{(\alpha)} + \sum_i \left[\mu_i^{(\beta)}\right] n_i^{(\beta)} + \sum_i \left[\mu_i^{(\gamma)}\right] n_i^{(\gamma)} \quad [2.16]$$

$$H = TS + \sigma A + \sum_i \left[\mu_i^{(\alpha)}\right] n_i^{(\alpha)} + \sum_i \left[\mu_i^{(\beta)}\right] n_i^{(\beta)} + \sum_i \left[\mu_i^{(\gamma)}\right] n_i^{(\gamma)} \quad [2.17]$$

$$F = -PV + \sigma A + \sum_i \left[\mu_i^{(\alpha)}\right] n_i^{(\alpha)} + \sum_i \left[\mu_i^{(\beta)}\right] n_i^{(\beta)} + \sum_i \left[\mu_i^{(\gamma)}\right] n_i^{(\gamma)} \quad [2.18]$$

$$G = \sigma A + \sum_i \left[\mu_i^{(\alpha)}\right] n_i^{(\alpha)} + \sum_i \left[\mu_i^{(\beta)}\right] n_i^{(\beta)} + \sum_i \left[\mu_i^{(\gamma)}\right] n_i^{(\gamma)} \quad [2.19]$$

$$G_\sigma = \sum_i \left[\mu_i^{(\alpha)}\right] n_i^{(\alpha)} + \sum_i \left[\mu_i^{(\beta)}\right] n_i^{(\beta)} + \sum_i \left[\mu_i^{(\gamma)}\right] n_i^{(\gamma)} \quad [2.20]$$

If we look again at relation [1.51], which remains valid in the case of a multicomponent system, then $F^{(\alpha)}$, which is the Helmholtz energy of a massive phase, is therefore a function of $T, V^{(\alpha)}, n_i^{(\alpha)}$. Similarly, $F^{(\beta)}$, which is also the internal energy of a massive phase, is also a function of $T, V^{(\beta)}, n_i^{(\beta)}$, and the global function F, which is a function of all of the variables, is a function of $T, V^{(\alpha)}, V^{(\beta)}, A, n_i^{(\alpha)}, n_i^{(\beta)}, n_i^{(\gamma)}$. We deduce from expression [1.51] that the internal Helmholtz energy of the surface phase $F^{(\gamma)}$ is a function of $T, V^{(\alpha)}, V^{(\beta)}, A, n_i^{(\alpha)}, n_i^{(\beta)}, n_i^{(\gamma)}$. This shows us that this phase is not autonomous. Therefore, we can write the partial differentials of the Helmholtz energy in the forms:

$$\frac{\partial F}{\partial n_i^{(\alpha)}} = \frac{\partial F^{(\alpha)}}{\partial n_i^{(\alpha)}} + \frac{\partial F^{(\gamma)}}{\partial n_i^{(\alpha)}} \quad [2.21a]$$

$$\frac{\partial F}{\partial n_i^{(\beta)}} = \frac{\partial F^{(\beta)}}{\partial n_i^{(\beta)}} + \frac{\partial F^{(\gamma)}}{\partial n_i^{(\beta)}} \quad [2.21b]$$

$$\frac{\partial F}{\partial n_i^{(\gamma)}} = \frac{\partial F^{(\gamma)}}{\partial n_i^{(\gamma)}} \qquad [2.21c]$$

As the partial differential $\frac{\partial F^{(\alpha)}}{\partial n_i^{(\alpha)}}$, for example, is the chemical potential of component i in the α phase, based on relation [2.14], we deduce the three relations:

$$\left[\mu_i^{(\alpha)}\right] = \mu_i^{(\alpha)} + \frac{\partial F^{(\gamma)}}{\partial n_i^{(\alpha)}} \qquad [2.22a]$$

$$\left[\mu_i^{(\beta)}\right] = \mu_i^{(\beta)} + \frac{\partial F^{(\gamma)}}{\partial n_i^{(\beta)}} \qquad [2.22b]$$

$$\left[\mu_i^{(\gamma)}\right] = \mu_i^{(\gamma)} \qquad [2.22c]$$

If we let $f^{(\gamma)}$ represent the Helmholtz energy of the surface layer per unit area (see relation [1.37]), using relations [1.37] and [2.1] we can also calculate:

$$\frac{\partial F}{\partial n_i^{(\gamma)}} = \frac{A \partial f^{(\gamma)}}{\partial \Gamma_i} = \mu_i^{(\gamma)} \qquad [2.23]$$

This last equation links the chemical potential of component i in the layer at the excess surface of i.

2.2.2. Chemical potentials and lateral chemical potentials

We shall now define values $\lambda_i^{(\alpha)}$ and $\lambda_i^{(\beta)}$, which are called the *lateral chemical potentials* of component i, by the equations:

$$\lambda_i^{(\alpha)} = \frac{\partial f^{(\gamma)}}{\partial C_i^{(\alpha)}} \qquad [2.24a]$$

and:

$$\lambda_i^{(\beta)} = \frac{\partial f^{(\gamma)}}{\partial C_i^{(\beta)}} \qquad [2.24b]$$

Thus, according to definitions [2.24a] and [2.24b], the values $\lambda_i^{(\alpha)}$, $\lambda_i^{(\beta)}$ and $\mu_i^{(\gamma)}$ are (according to relation [2.23]) functions of the intensive variables (T, $C_i^{(\alpha)}, C_i^{(\beta)}$, Γ_i). Hence, they are intensive values.

By derivation in relation to the quantity of component i in α phase we can write:

$$\frac{\partial F^{(\gamma)}}{\partial n_i^{(\alpha)}} = A\frac{\partial f^{(\gamma)}}{\partial n_i^{(\alpha)}} = \frac{A}{V^{(\alpha)}}\frac{\partial F^{(\gamma)}}{\partial C_i^{(\alpha)}} = \frac{A}{V}\lambda_i^{(\alpha)} \qquad [2.25]$$

Similarly, for the quantity of component i in the β phase:

$$\frac{\partial F^{(\gamma)}}{\partial n_i^{(\beta)}} = \frac{A}{V}\lambda_i^{(\beta)} \qquad [2.26]$$

The values $\lambda_i^{(\alpha)}$ and $\lambda_i^{(\beta)}$ are known as the *lateral chemical potentials of component i*, reflecting the fact that they give the contribution to the Helmholtz energy of the concentration of i on both sides of the surface.

NOTE.– We can see, from definition [2.24], that for pure substances the lateral chemical potential is zero.

By feeding back equations [2.25] and [2.26] into relations [2.24a] and [2.24b], we obtain:

$$\left[\mu_i^{(\alpha)}\right] = \mu_i^{(\alpha)} + \frac{A}{V^{(\alpha)}}\lambda_i^{(\alpha)} \qquad [2.27a]$$

$$\left[\mu_i^{(\beta)}\right] = \mu_i^{(\beta)} + \frac{A}{V^{(\beta)}}\lambda_i^{(\beta)} \qquad [2.27b]$$

$$\left[\mu_i^{(\gamma)}\right] = \mu_i^{(\gamma)} \qquad [2.27c]$$

These relations demonstrate that the complete chemical potentials $\left[\mu_i^{(\alpha)}\right]$ and $\left[\mu_i^{(\beta)}\right]$ are no longer intensive variables because if the chemical potentials and the lateral chemical potentials are intensive, the involvement of the geometric ratios A/V makes the complete chemical potentials of the bulk phases dependent on the amount of material, amongst others.

2.2.3. *Conditions of equilibrium in a capillary system*

The general equilibrium condition for the three-phase system is expressed by the equality of the generalized chemical potentials – in our case the complete chemical potentials – of each of the three phases:

$$\left[\mu_i^{(\alpha)}\right] = \left[\mu_i^{(\beta)}\right] = \left[\mu_i^{(\gamma)}\right] \qquad [2.28]$$

In view of relations [2.27a], [2.27b] and [2.27c], this double equality becomes:

$$\mu_i^{(\alpha)} + \frac{A}{V^{(\alpha)}}\lambda_i^{(\alpha)} = \mu_i^{(\beta)} + \frac{A}{V^{(\beta)}}\lambda_i^{(\beta)} = \mu_i^{(\gamma)} \qquad [2.29]$$

We shall accept, as do Defay and Prigogine, that any portion of a system at equilibrium is also at equilibrium, which is tantamount to accepting that the equilibrium conditions must be able to be expressed on the basis of the intensive values alone. However, we have seen that the complete chemical potentials were not intensive values because of the terms A/V.

Let us represent our system by using two volumic phases with the volumes $V^{(\alpha)}$ and $V^{(\beta)}$ and an interface Σ (Figure 2.2). Our system ABCD can be mutilated by cutting off portion CDEF, and be reduced to ABEF without changing the equilibrium conditions. Then, only the term $V^{(\beta)}$ is reduced and, with the equilibrium conditions being the same, which means that the terms containing this volume in the complete potentials must assume the value of 0, at equilibrium.

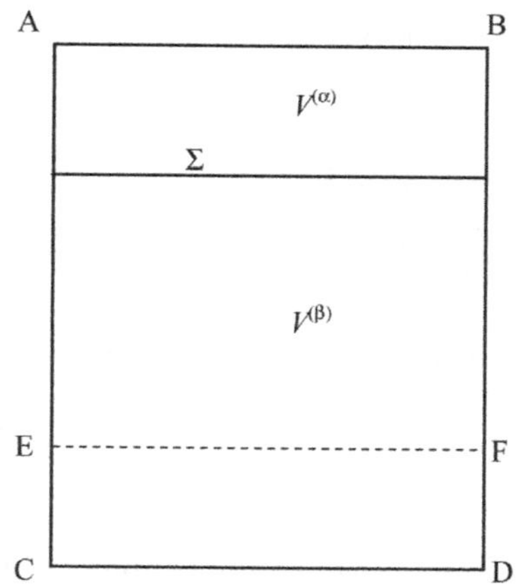

Figure 2.2. *Independence of the conditions of equilibrium and the dimensions of the system*

This observation is expressed by the zeroing of the lateral chemical potentials at equilibrium:

$$\lambda_i^{(\alpha)} = \lambda_i^{(\beta)} = 0 \qquad [2.30]$$

Finally, in view of relations [2.27a], [2.27b] and [2.27c], the equilibrium condition of the system is expressed by the equality of the *ordinary* chemical potentials:

$$\mu_i^{(\alpha)} = \mu_i^{(\beta)} = \mu_i^{(\gamma)} \qquad [2.31]$$

The equilibrium condition of a capillary system (containing an interface) is expressed, as for a volume system, by the equality of the chemical potentials in the three phases: the two bulk phases and the surface phase.

2.2.4. *Gibbs–Duhem relation for surface phenomena*

Look again at expression [2.20] of the generalized Gibbs energy function. By differentiation, we obtain:

$$\mathrm{d}G_\sigma = \sum_i \mu_i \,\mathrm{d}n_i + \sum_i n_i \,\mathrm{d}\mu_i \qquad [2.32]$$

However, the transposition of relation [2.32] to the case of polycomponent systems gives us:

$$\mathrm{d}G_\sigma = -S^{(\alpha)}\,\mathrm{d}T - S^{(\beta)}\,\mathrm{d}T - S^{(\gamma)}\,\mathrm{d}T + V^{(\alpha)}\,\mathrm{d}P^{(\alpha)} + V^{(\beta)}\,\mathrm{d}P^{(\beta)} - A\,\mathrm{d}\sigma + \sum \mu_i\,\mathrm{d}n_i \qquad [2.33]$$

By combining relations [2.32] and [2.33], we find:

$$S^{(\alpha)}\,\mathrm{d}T + S^{(\beta)}\,\mathrm{d}T + S^{(\gamma)}\,\mathrm{d}T - V^{(\alpha)}\,\mathrm{d}P^{(\alpha)} - V^{(\beta)}\,\mathrm{d}P^{(\beta)} + A\,\mathrm{d}\sigma + \sum_i n_i\,d\mu_i = 0 \qquad [2.34]$$

This expression is the transposition of the Gibbs–Duhem relation to surface systems. In particular, at constant temperature, pressure and surface tension it is expressed in the form:

$$\sum_i n_i\,d\mu_i = 0 \qquad [2.35]$$

This relation is absolutely identical to the Gibbs–Duhem relation obtained for the bulk phases at constant pressure and temperature.

2.2.5. *Adsorption and Gibbs isotherm*

Let us apply the Gibbs–Duhem relation to all the two volumic α and β phases of the layered model. We have:

$$S^{(\alpha)}\,\mathrm{d}T + S^{(\beta)}\,\mathrm{d}T - V^{(\alpha)}\,\mathrm{d}P^{(\alpha)} - V^{(\beta)}\,\mathrm{d}P^{(\beta)} + \sum_i \left(n_i^{(\alpha)} + n_i^{(\beta)}\right) d\mu_i = 0 \qquad [2.36]$$

By subtracting equation [2.36] from equation [2.34], taking account of relation [2.2], we find:

$$S^{(\gamma)}\,\mathrm{d}T + A\,\mathrm{d}\sigma + \sum_i n_i^{(\gamma)}\,d\mu_i = 0 \qquad [2.37]$$

or, by dividing by the area A:

$$\mathrm{d}\sigma = -s^{(\gamma)}\,\mathrm{d}T - \sum_i \Gamma_i\,d\mu_i \qquad [2.38]$$

In this expression, the volumes no longer play a part, and at constant temperature, equation [2.38] is simplified to:

$$d\sigma = -\sum_i \Gamma_i d\mu_i \qquad [2.39]$$

The result of this is that the excess concentration of component *i* is:

$$\Gamma_i = -\left(\frac{\partial \sigma}{\partial \mu_i}\right)_{T,P,\mu_j} \qquad [2.40]$$

This relation constitutes the *first form of the Gibbs isotherm of adsorption.*

This expression has the peculiarity of being impossible to use, because the chemical potentials, which are linked by the Gibbs–Duhem relation, are not independent. Therefore, it is impossible to differentiate relation [2.41] in relation to one of the chemical potentials whilst keeping all of the others constant. We shall circumvent this difficulty by replacing the absolute adsorption of component *i* by its relative adsorption in relation to a chosen component – say, component 1 – defined by relation [2.7], recapped here:

$$\Gamma_{i,1} = \Gamma_i - \Gamma_1 \frac{C_i^{(\alpha)} - C_i^{(\beta)}}{C_1^{(\alpha)} - C_1^{(\beta)}} \qquad [2.41]$$

For this purpose, we first note that the quantities of material in the two bulk phases must obey the Gibbs–Duhem relation for the bulk phases:

$$\sum_i n_i^{(\alpha)}\, d\mu_i = \sum_i n_i^{(\beta)}\, d\mu_i = 0 \qquad [2.42]$$

These relations lead to the concentrations:

$$\sum_i C_i^{(\alpha)}\, d\mu_i = \sum_i C_i^{(\beta)}\, d\mu_i = 0 \qquad [2.43]$$

and thus we have:

$$\sum_i \Gamma_i \, \mathrm{d}\mu_i = \sum_i \Gamma_{i,1} \, \mathrm{d}\mu_i = 0 \qquad [2.44]$$

and we find the relation:

$$\mathrm{d}\sigma = -\sum_i \Gamma_{i,1} d\mu_i \qquad [2.45]$$

In this relation, because $\Gamma_{1,1}$ has the value of 0, we merely need to perform the calculation on the basis of $i = 2$. Thus, we have:

$$\Gamma_{i,1} = -\left(\frac{\partial \sigma}{\partial \mu_i}\right)_{T,P,\mu_j} \quad \text{where } i \neq 1, \text{ and } i \geq 2 \qquad [2.46]$$

We have therefore avoided the problem, because in this expression differentiation is possible because, as the potential μ_1 plays no part, the potentials μ_j are all independent.

If the solution is ideal, the chemical potential of the component i is:

$$\mu_i = \mu_i^0 + \mathrm{R}T \ln C_i \qquad [2.47]$$

Relation [2.46] then takes the form:

$$\Gamma_{i,1} = -\frac{C_i}{\mathrm{R}T}\left(\frac{\partial \sigma}{\partial C_i}\right)_{T,P,C_j} \qquad [2.48]$$

This relation is the *second form of the Gibbs isotherm of adsorption*. It enables us to determine the relative adsorptions of the different components by varying their concentrations. It shows that it is the components tending to decrease the surface tension which are adsorbed positively ($\Gamma_{i,1} > 0$).

2.3. Surface tension of solutions

The thermodynamic results established above do not depend on the nature of the α and β phases present, which may be solid, liquid or gaseous.

We shall now turn our attention to the properties of the interfaces between a liquid solution and its own vapor phase at equilibrium. In keeping with all of the above, the vapor phase will be the β phase, with the α phase then being the liquid solution, whose nature we shall not specify further.

As the β phase is gaseous, its Helmholtz energy, Gibbs energy and generalized Gibbs energy in relation to the surface phenomena will be negligible in relation to those of the liquid phase, and therefore the overall value in the system is the sum of the corresponding contributions of the liquid phase and the layer. Thus, relation [1.34] for the generalized Gibbs energy of the layer (γ phase) gives us:

$$G_{\sigma}^{(\gamma)} = G^{(\gamma)} - \sigma A \qquad [2.49]$$

By differentiating and remembering that the differential of the function *G* is the chemical potential, we obtain the following, in light of relation [2.15]:

$$\left(\frac{\partial G_{\sigma}^{(\gamma)}}{\partial n_i^{(\gamma)}}\right)_{T,\sigma,n_j^{(\gamma)}} = \mu_i^{(\gamma)} + \sigma\left(\frac{\partial A}{\partial n_i^{(\gamma)}}\right)_{T,\sigma,n_j^{(\gamma)}} = \left[\mu_i^{(\gamma)}\right] \qquad [2.50]$$

By explicitly stating the chemical potential as a function of the activity, we obtain:

$$\left[\mu_i^{(\gamma)}\right] = \mu_i^{0(\gamma)} + RT\ln a_i^{(\gamma)} - \sigma\frac{\partial A}{\partial n_i^{(\gamma)}} \qquad [2.51]$$

However, at equilibrium, the chemical potentials are equal, so:

$$\left[\mu_i^{(\gamma)}\right] = \mu_i^{(\gamma)} = \mu_i \qquad [2.52]$$

The chemical potential of the component *i* in the liquid is of the form:

$$\mu_i = g_i^0 + RT\ln a_i \qquad [2.53]$$

By substituting expressions [2.52] and [2.53] back into relation [2.51], we obtain:

$$g_i^{0(\gamma)} + RT \ln a_i^{(\gamma)} - \sigma \frac{\partial A}{\partial n_i^{(\gamma)}} = g_i^0 + RT \ln a_i \qquad [2.54]$$

This formula links the activity, and therefore the composition, of the surface layer at equilibrium with the liquid solution to its properties.

Note that the standard terms $g_i^{0(\gamma)}$ and g_i^0 are functions only of the temperature and, a little (in the condensed phase), on the pressure.

NOTE.– We would employ the same reasoning and obtain the same results if we were to replace the liquid solution with a solid solution in the presence of its vapor. Hence, condition [2.54] is the equilibrium condition of a condensed solution (liquid or solid) and of its surface in the presence of its vapor.

2.3.1. *Perfect solutions*

In the case of perfect solutions, for which the activity coefficients are equal to 1, relation [2.54] becomes:

$$g_i^{0(\gamma)} + RT \ln x_i^{(\gamma)} - \sigma \frac{\partial A}{\partial n_i^{(\gamma)}} = g_i^0 + RT \ln x_i \qquad [2.55]$$

In addition, we know that we can model these solutions by choosing very similar dimensions of the molecules of the different components, which can be expressed by the introduction of a common term A_M, such that:

$$\frac{\partial A}{\partial n_1^{(\gamma)}} = \frac{\partial A}{\partial n_2^{(\gamma)}} = \frac{\partial A}{\partial n_3^{(\gamma)}} \ldots = A_M \qquad [2.56]$$

The term A_M represents the molar area, and its value can be estimated on the basis of the molar volumes (or the densities) as follows:

Assimilate the molecule to a small cube whose volume is V_m/Na. The area occupied by a molecule would be:

$$A_m = \left(\frac{V_m}{\mathrm{N_a}} \right)^{2/3} \qquad [2.57]$$

Thus, the molar area is:

$$A_M = \mathrm{N_a} A_m = V_m^{2/3} \mathrm{N_a^{1/3}} = \left(\frac{M}{\rho} \right)^{2/3} \mathrm{N_a^{1/3}} \qquad [2.58]$$

Now consider the pure component i; by applying relation [2.55], we would have:

$$g_i^{0(\gamma)} - g_i^0 = \sigma_i A_M \qquad [2.59]$$

where σ_i is the surface tension of the pure liquid i in the presence of its vapor.

By feeding this relation into equation [2.55], we obtain the following for the surface tension of the solution:

$$\sigma = \sigma_i + \frac{\mathrm{R}T}{A_M} \ln \frac{x_i^{(\gamma)}}{x_i} \qquad [2.60]$$

Now consider a binary system formed of components 1 and 2. We eliminate the term σ between the two relations [2.60] expressed for each of these components. We obtain:

$$\frac{x_1^{(\gamma)}}{x_2^{(\gamma)}} = \frac{x_1 \exp - \dfrac{\sigma_1 A_M}{\mathrm{R}T}}{x_2 \exp - \dfrac{\sigma_2 A_M}{\mathrm{R}T}} \qquad [2.61]$$

Thus, for component 1:

$$x_1^{(\gamma)} = \frac{x_1 \exp - \dfrac{\sigma_1 A_M}{\mathrm{R}T}}{x_1 \exp - \dfrac{\sigma_1 A_M}{\mathrm{R}T} + x_2 \exp - \dfrac{\sigma_2 A_M}{\mathrm{R}T}} \qquad [2.62a]$$

and for component 2:

$$x_2^{(\gamma)} = \frac{x_2 \exp - \dfrac{\sigma_2 A_M}{RT}}{x_1 \exp - \dfrac{\sigma_1 A_M}{RT} + x_2 \exp - \dfrac{\sigma_2 A_M}{RT}} \quad [2.62b]$$

These are the *Schuchowitsky relations*.

In light of the fact that $x_1^{(\gamma)} + x_2^{(\gamma)} = 1$, we can deduce, using relation [2.60], that:

$$\exp - \frac{\sigma A_M}{RT} = x_1 \exp - \frac{\sigma_1 A_M}{RT} + x_2 \exp - \frac{\sigma_2 A_M}{RT} \quad [2.63]$$

Let us calculate the molar fraction in the layer. We obtain:

$$\sigma_1 - \sigma = \frac{RT}{A_M} \ln \left\{ 1 + x_2 \left[\exp \left\{ \frac{(\sigma_1 - \sigma_2) A_M}{RT} \right\} - 1 \right] \right\} \quad [2.64]$$

Relations [2.64], applied to each of the two components, were put forward by Szyszkowski, on the basis of experimental results, in the form:

$$\sigma_1 - \sigma = b \ln \left(1 + \frac{C_2}{a} \right) \quad [2.65]$$

with the values a and b being two constants independent of the compositions.

If the surface tensions of the two components in the solution are fairly close, we can make the hypothesis:

$$\left| \frac{(\sigma_1 - \sigma_2) A_M}{RT} \right| << 1$$

Then, by developing the serial exponential, then the logarithm, relation [2.62] is simplified to become:

$$\sigma = \sigma_1 x_1 + \sigma_2 x_2 \tag{2.66}$$

Thus, the surface tension of a perfect solution, as a first approximation, obeys a law of additivity as a function of the surface tensions of its pure components.

2.3.2. *Highly-dilute solutions*

For the dilute solutions, with a solvent whose molar fraction is x_0 and a solute whose molar fraction is x_s, the chemical potentials obey relation [2.53]. The activity coefficients are equal to 1 in the case of the solvent and, in the case of the solute, to a reference chemical potential μ_s^∞ (the chemical potential in the infinitely-dilute solution), which is a function only of the temperature and pressure.

Applying relation [2.51] to the pure solvent, we obtain:

$$\mu_0 = g_0^{(\gamma)} - \sigma_0 A_{M0} \tag{2.67}$$

and thus, relation [2.60] gives us:

$$(\sigma - \sigma_0) A_{M0} = RT \ln \frac{x_0^{(\gamma)}}{x_0} \tag{2.68}$$

For the solute s, we can apply relation [2.55], which is then written as:

$$\mu_s^{\infty(\gamma)} + RT \ln x_s^{(\gamma)} - \sigma A_{Ms} = \mu_s^\infty + RT \ln x_s \tag{2.69}$$

By eliminating the term σ between relations [2.68] and [2.69], we obtain:

$$\frac{RT}{A_{Ms}} \ln \frac{x_s^{(\gamma)}}{x_s} - \frac{RT}{A_{M00}} \ln \frac{x_0^{(\gamma)}}{x_0} = \sigma_0 - \frac{\mu_s^{\infty(\gamma)} - \mu_s^\infty}{A_{Ms}} \tag{2.70}$$

For a very dilute solution, we have: $x_s^{(\gamma)} \ll 1$ and $x_s \ll 1$. The second logarithm in relation [2.70] is negligible in comparison to the first, and we can write:

$$\frac{x_s^{(\gamma)}}{x_s} \cong \exp\frac{\sigma_0 A_{Ms} - \left(\mu_s^{\infty(\gamma)} - \mu_s^{\infty}\right)}{RT} \qquad [2.71]$$

The right-hand side of this expression is an adsorption energy, which depends only on the temperature and pressure. Thus, in a highly-dilute solution, the molar fraction of the solute in the layer is proportional to its molar fraction in the solution.

By feeding back into expression [2.68], we find:

$$(\sigma - \sigma_0) A_{M0} \cong RT\left(x_s - x_s^{(\gamma)}\right) = RTx_s\left\{1 - \exp\frac{\left[\sigma_0 A_{Ms} - \left(\mu_s^{\infty(\gamma)} - \mu_s^{\infty}\right)\right]}{RT}\right\} \qquad [2.72]$$

The surface tension of a sufficiently-dilute solution varies in linear fashion with the molar fraction of the solute. Thus, we find the practical formula given by Traube, who formulated that variation in the form:

$$\sigma = \sigma_0 - bx_s \qquad [2.73]$$

NOTE.– The Szyszkowski relation [2.65] becomes identical to the Traube relation [2.73] if the molar fraction of one of the components tends toward zero.

2.4. Interface tension between two liquids

When two immiscible liquids are brought into contact, the separating surface, known as the interface, has the same properties as the surface of a liquid in contact with its vapor. The thermodynamics of the interface are dealt with in the same way, so we see the existence of a surface tension which, here, is called the *interfacial tension*.

Attempts have been made to establish a relation between the interfacial tension between two liquids, A and B, and the surface energies of each of the liquids. We can see through experimentation that the interfacial tension is always less than the surface energy of whichever pure liquid has the highest surface energy. Antonov proposed a simple rule expressing the fact that the interface tension was the absolute value of the difference between the surface energies of the two liquids:

$$\sigma_{AB} = |\sigma_A - \sigma_B| \qquad [2.74]$$

In reality, if the two liquids are partially miscible, it is the difference between the surface energies of the saturated solutions which must be considered.

We can see experimentally that, whilst the interfacial tension may be greater than that predicted by Antonov's rule, it is never less.

The interfacial tension is lower when the solubility of the liquids in one another is higher. If the two liquids are of the same nature, the interfacial tension becomes zero when the two liquids become miscible, meaning when the two phases come together at the critical temperature.

The condition for complete miscibility of two liquids is therefore that their interfacial tension be null (or negative).

2.5. Energy of adhesion of two liquids

Consider two liquids A and B, brought into contact with one another in a column with a surface of 1 unit. We use the term *energy of adhesion* to describe the work necessary to separate the surfaces of the two liquids in contact.

When we imagine this separation to be performed, we remove the interface of unit area and create two surfaces of unit area for each of the liquids. The creation of these two free surfaces requires an amount of work equal to the sum $\sigma_A + \sigma_B$, whereas the removal of the interface produces work equal to σ_{AB}. The adhesion energy W_{AB} will therefore be:

$$W_{AB} = \sigma_A + \sigma_B - \sigma_{AB} \qquad [2.75]$$

This is the *Dupré relation.* W_{AB} is expressed in joules/m^2.

If the two liquids are partially miscible, it is the surface energies of the two saturated solutions that must be used in Dupré's relation instead of the surface tensions of the two pure liquids.

2.6. Spreading of a liquid over another liquid

Now suppose that we deposit a drop of liquid A on the surface of liquid B which is immiscible with A. Liquid A may either remain in the state of a drop on top of B, or spread over B to form a film. In the latter case, we say that *liquid A wets liquid B.*

We shall now try to find the conditions of wetting of B by A – i.e. the conditions for liquid A to spread in a film over the surface of liquid B. If liquid A remains as a drop, it is possible to determine the shape of that drop.

In order to express the equilibrium condition of such a system, let us assimilate the surface energies to force-vectors tangential to the surfaces at a point on the line with three phases, or the triple line. To do this, we choose a vertical cross-section of the drop passing through its axis of symmetry, which reveals a triple point where we apply the vectors (Figure 2.3(a)). We let θ_A denote the angle between the forces σ_B and σ_{AB}. The definition of the other angle θ_B is similar between the forces σ_A and σ_{AB}. The equilibrium condition is expressed on the basis of the Neumann triangle (Figure 2.3(b)) by the relation:

$$\frac{\sigma_A}{\sin\theta_B} = \frac{\sigma_B}{\sin\theta_A} = \frac{\sigma_{AB}}{\sin(2\pi - \theta_A - \theta_B)} \qquad [2.76]$$

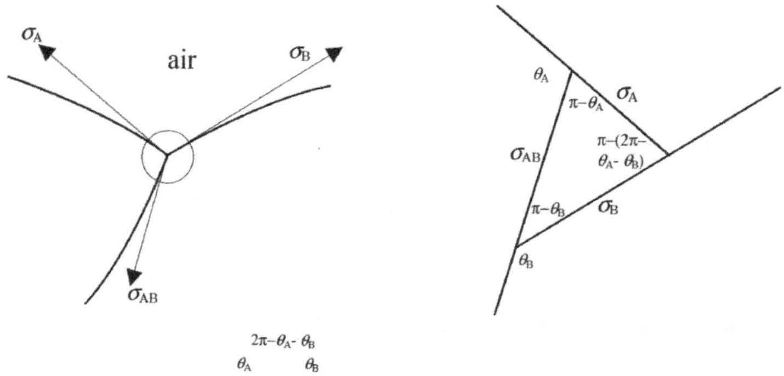

a) b)

Figure 2.3. *Spreading of a drop over a liquid: a) equilibrium of the triple line; b) the Neumann triangle*

This condition can only be satisfied if the greatest surface energy is less than the sum of the other two. We know that the interfacial energy σ_{AB} is less than the greatest surface energy. The equilibrium in question is impossible if we have:

$$\sigma_B \geq \sigma_A + \sigma_{AB} \qquad [2.77]$$

as σ_B is supposed to be greater than σ_A. For given values of σ_A and σ_{AB}, the angle θ_A decreases if σ_B increases. It is zero when σ_B reaches the value $\sigma_A + \sigma_{AB}$. In this case, there is the spreading of liquid A over the surface of B. The same is true, *a fortiori*, if σ_B continues to grow.

The spreading condition can also be formulated as:

$$\sigma_B - \sigma_{AB} \geq \sigma_A \qquad [2.78]$$

If we take account of relation [2.75], this condition can also be written:

$$W_{AB} \geq 2\sigma_A \qquad [2.79]$$

where $2\sigma_A$ is the *cohesion energy* of liquid A. Thus, the condition of spreading of liquid A over liquid B is that the energy of adhesion of liquid A to liquid B be at least equal to the cohesion energy of liquid A.

Harkins used the term *spreading coefficient* of liquid A over liquid B to describe the difference between the adhesion energy and the cohesion energy:

$$h_{spr} = W_{\mathrm{AB}} - 2\sigma_{\mathrm{A}} = \sigma_{\mathrm{B}} - \sigma_{\mathrm{A}} - \sigma_{\mathrm{AB}} \qquad [2.80]$$

We see the spreading of one liquid over another when its spreading coefficient is positive or null. The liquid does not spread if the spreading coefficient is negative. Harkins determined a number of spreading coefficients of liquids over water at 20°C. The values found range from 0.049 $\mathrm{J.m^{-2}}$ for N-propyl alcohol to –0.026 $\mathrm{J.m^{-2}}$ for methylene iodide. These coefficients increase or decrease with temperature, and we may see a switch in the sign of the coefficient as the temperature varies.

NOTE.– The aforementioned values are valid for rigorously clean surfaces, exempt from all traces of impurities, which it is extremely difficult to obtain.

If liquid A spreads over liquid B, it is impossible for liquid B to spread over liquid A. Indeed, for that to be possible, we would have to simultaneously satisfy the two conditions:

$$\sigma_{\mathrm{B}} - \sigma_{\mathrm{A}} - \sigma_{\mathrm{AB}} \geq 0 \qquad [2.81]$$

$$\sigma_{\mathrm{A}} - \sigma_{\mathrm{B}} - \sigma_{\mathrm{AB}} \geq 0 \qquad [2.82]$$

By adding together those two conditions, we find:

$$\sigma_{\mathrm{AB}} \leq 0 \qquad [2.83]$$

This means that the two liquids must be totally intermiscible, which runs counter to our fundamental hypothesis.

We can now examine how Antonov's rule (relation [2.74]) fits in with the spreading condition. If $\sigma_{\mathrm{B}} > \sigma_{\mathrm{A}}$, the condition for two liquids not to spread over one another is:

$$\sigma_{\mathrm{B}} < \sigma_{\mathrm{A}} + \sigma_{\mathrm{AB}} \qquad [2.84]$$

Thus:

$$\sigma_{\mathrm{AB}} > \sigma_{\mathrm{B}} - \sigma_{\mathrm{A}} \qquad [2.85]$$

and Antonov's formula [2.74] is obviously not satisfied.

If a liquid can be spread over another – e.g. A over B – then, in accordance with relation [2.77], we have:

$$\sigma_B \geq \sigma_A + \sigma_{AB} \quad [2.86]$$

However, in light of the large contact surface, there is slight dissolution of the liquids in one another, so that their surface energies have become σ'_A and σ'_B, and if Antonov's rule is so often satisfied, this means that we have:

$$\sigma'_B = \sigma'_A + \sigma_{AB} \quad [2.87]$$

This case is encountered for the water/benzene couple. The spreading coefficient, which is 0.0098 for pure liquids, takes the value of 0 when each liquid is saturated by the other.

2.7. Example of the microscopic modeling of surfaces of solutions: the monolayer model for strictly-regular solutions

A simple model, in the case of the surfaces of strictly-regular solutions, can help us to understand a number of experimental phenomena.

2.7.1. *Presentation of the model*

We can model the surface of a liquid using the monolayer model introduced in section 1.7. We retain the hypothesis of potential interactions between two molecules decreasing very rapidly, meaning we can stick to an interaction with near neighbors only, which, for a molecule in a monomolecular layer are divided into *lz* neighbors in the same layer and *mz* neighbors in each of the adjacent layers. The molecules of the gaseous phase are too far removed from the surface for their interactions to be taken into account. The solutions will be supposed to be strictly regular, and thus we accept that the molecules of components A and B have practically the same dimensions, which will enable us to preserve relation [2.56] and the excess entropy in the phases will be zero.

Let us define the system as a cylinder containing the surface layer, delimited on one side by the plane CD, which separates the solution and the surface layer, and on the other side by a line GH that is parallel to the surface, which delimits the volume of the solution. Let *j* be the number of monomolecular layers between the planes CD and GH within the liquid phase (Figure 2.4).

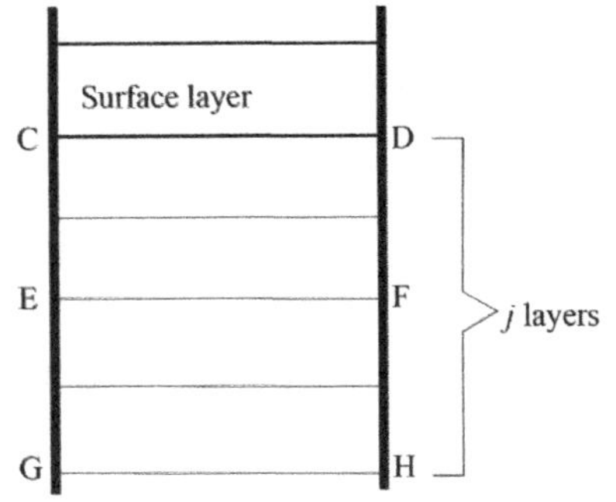

Figure 2.4. *The molecular layers in the volume and on the surface of a liquid*

Each molecule in a layer has *zl* near neighbors in the same sheet and *zm* neighbors in the next monomolecular layer.

The total number of molecules in the surface layer is $N_a n^{(\gamma)}$ with the respective molar fractions of $x_1^{(\gamma)}$ and $x_2^{(\gamma)}$. The molar fractions within the liquid are x_1 and x_2, with the numbers of molecules being $N_a n_1$ and $N_a n_2$ (N_a is Avogadro's number).

Let us list the different couples.

Across the CD plane, therefore, we shall have, on average:

– $N_a n^{(\gamma)} x_1^{(\gamma)} zmx_1$ couples 1-1, with energy ε_{11};

– $N_a n^{(\gamma)} x_2^{(\gamma)} zmx_2$ couples 2-2, with energy ε_{22};

– $N_a n^{(\gamma)} x_1^{(\gamma)} zmx_2 + N_a n^{(\gamma)} x_2^{(\gamma)} zmx_1$ couples 1-2, with energy ε_{12}.

In the surface layer, we have:

– $N_a n^{(\gamma)} zl\left(x_1^{(\gamma)}\right)^2$ couples 1-1, with energy ε_{11};

– $N_a n^{(\gamma)} zl\left(x_2^{(\gamma)}\right)^2$ couples 2-2, with energy ε_{22};

– $2N_a n^{(\gamma)} zl\left(x_1^{(\gamma)}\right) x_2^{(\gamma)}$ couples 1-2, with energy ε_{12}.

Based on this list, we shall evaluate the Gibbs energy of the system and the chemical potentials.

2.7.2. *Chemical potentials of the surface and bulk components of a strictly-regular solution*

The portion of the internal energy due to the interactions between two layers is:

$$U_{\text{mixed}} = \mathrm{N_a} n^{(\gamma)} zm \left[x_1^{(\gamma)} x_1 \varepsilon_{11} + x_2^{(\gamma)} x_2 \varepsilon_{22} + \left(x_1^{(\gamma)} x_2 + x_2^{(\gamma)} x_1 \right) \varepsilon_{12} \right] \qquad [2.88]$$

The contribution of the interactions within the surface layer is:

$$U_{\text{layer}} = \frac{\mathrm{N_a} n^{(\gamma)} zl}{2} \left[\left(x_1^{(\gamma)} \right)^2 \varepsilon_{11} + \left(x_2^{(\gamma)} \right)^2 \varepsilon_{22} + 2 \left(x_1^{(\gamma)} x_2^{(\gamma)} \right) \varepsilon_{12} \right] \qquad [2.89]$$

In the same way, we evaluate the contribution of the molecules situated in a layer within the liquid, in expressions [2.88] and [2.89] we simply need to replace $x_1^{(\gamma)}$ and $x_2^{(\gamma)}$ with x_1 and x_2 respectively. The expression thus obtained must be multiplied by (j–1/2), because there are j planes such as EF. However, at the level of GH, in the application of relation [2.88], half of the bonds must be attributed to our system delimited by GH and the other half to the outer system.

By bringing into play the exchange energy w_{12} defined by the relation:

$$w_{12} = z \left(\varepsilon_{12} - \frac{\varepsilon_{11} + \varepsilon_{22}}{2} \right) \qquad [2.90]$$

add together the different contributions. By grouping together the terms, we find:

$$\begin{aligned} U_{\text{surf}} = {} & \frac{\mathrm{N_a} (n_1 + n_2) z}{2} (x_1 \varepsilon_{11} + x_2 \varepsilon_{22}) + \mathrm{N_a} z (n_1 + n_2) w_{12} x_1 x_2 \\ & + \frac{\mathrm{N_a} n_1^{(\gamma)} zl}{2} \left(x_1^{(\gamma)} \varepsilon_{11} + x_2^{(\gamma)} \varepsilon_{22} \right) + \mathrm{N_a} z n^{(\gamma)} w_{12} l x_1^{(\gamma)} x_2^{(\gamma)} \\ & + \frac{\mathrm{N_a} n_1^{(\gamma)} zm}{2} \left(x_1^{(\gamma)} \varepsilon_{11} + x_2^{(\gamma)} \varepsilon_{22} \right) + \mathrm{N_a} n^{(\gamma)} w_{12} m \left(x_1^{(\gamma)} x_2 + x_2^{(\gamma)} x_1 - x_1 x_2 \right) \end{aligned} \qquad [2.91]$$

In this sum, the right-hand side contains three lines:

– The first line, with two terms, represents the contribution of the interactions between molecules within the liquid bulk phase. These terms contain no quantity relating to the surface phase.

– The second line, with two terms, which contains only the quantities of material and the molar fractions of the surface layer, is the contribution of the interactions between molecules situated in the surface layer.

– The third line, which also contains two terms, is the contribution of the interactions between the molecules of the surface layer and those of the adjacent layer belonging to the bulk phase.

It follows that the term of the internal energy due to the surface layer is given by the sum of the last two lines:

$$\begin{aligned} U_{\text{surf}}^{(\gamma)} &= \frac{\mathrm{N_a} n_1^{(\gamma)} z(l+m)}{2}\left(x_1^{(\gamma)}\varepsilon_{11} + x_2^{(\gamma)}\varepsilon_{22}\right) \\ &+ \mathrm{N_a} z n^{(\gamma)} w_{12} l x_1^{(\gamma)} x_2^{(\gamma)} + \mathrm{N_a} n^{(\gamma)} w_{12} m\left(x_1^{(\gamma)} x_2 + x_2^{(\gamma)} x_1 - x_1 x_2\right) \end{aligned} \quad [2.92]$$

We then see that we can write:

$$U_{\text{surf}}^{(\gamma)} = U_{\text{surf}}^{(\gamma)*} + U_{\text{surf}}^{(\gamma)xs} \quad [2.93]$$

$U_{\text{surf}}^{(\gamma)*}$ is the internal energy of the layer if the solution is perfect ($w_{12} = 0$). This is our reference state and $U_{\text{surf}}^{(\gamma)xs}$ is the excess energy of the layer, given by:

$$\begin{aligned} U_{\text{surf}}^{(\gamma)xs} &= \mathrm{N_a} z n^{(\gamma)} w_{12} l x_1^{(\gamma)} x_2^{(\gamma)} \\ &+ \mathrm{N_a} n^{(\gamma)} w_{12} m\left(x_1^{(\gamma)} x_2 + x_2^{(\gamma)} x_1 - x_1 x_2\right) \end{aligned} \quad [2.94]$$

This expression can also be written in the form:

$$\begin{aligned} U_{\text{surf}}^{(\gamma)xs} &= \mathrm{N_a} z n^{(\gamma)} w_{12} l\left[x_1^{(\gamma)}\left(x_2^{(\gamma)}\right)^2 + x_2^{(\gamma)}\left(x_1^{(\gamma)}\right)^2\right] \\ &+ \mathrm{N_a} n^{(\gamma)} w_{12} m\left[x_1^{(\gamma)}\left(x_2^{(\gamma)}\right)^2 + x_2^{(\gamma)}\left(x_1^{(\gamma)}\right)^2\right] \end{aligned} \quad [2.95]$$

As regards the entropy of our system, we accept that the two surface and bulk phases have the same structure and that the solutions therein are strictly regular, meaning that the entropy is composed solely of the mixing term (the excess entropy is zero). The total entropy of the system would be:

$$S = S^{(\alpha)} + S^{(\gamma)} = x_1 \ln x_1 + x_2 \ln x_2 + x_1^{(\gamma)} \ln x_1^{(\gamma)} + x_2^{(\gamma)} \ln x_2^{(\gamma)} \quad [2.96]$$

The Helmholtz energy, which is also the Gibbs energy (the layer has no term PV), of our layer is then:

$$G^{(\gamma)} = U_{\text{surf}}^{(\gamma)*} + U_{\text{surf}}^{(\gamma)xs} - TS^{(\gamma)} = G^{(\gamma)*} + U_{\text{surf}}^{(\gamma)xs} \quad [2.97]$$

where $G^{(\gamma)*}$ is the part of the Helmholtz energy of a perfect solution.

In light of equations [2.95] and [2.96], relation [2.97] is written:

$$\begin{aligned} G^{(\gamma)} &= n_1^{(\gamma)}\left[\text{g}_1^{0(\gamma)} + \text{R}T \ln x_1^{(\gamma)} + \text{N}_\text{a} z w_{12} l \left(x_2^{(\gamma)}\right)^2 \right] \\ &+ n_2^{(\gamma)}\left[\text{g}_2^{0(\gamma)} + \text{R}T \ln x_2^{(\gamma)} + \text{N}_\text{a} z w_{12} l \left(x_2^{(\gamma)}\right)^2 \right] \\ &+ \text{N}_\text{a} z w_{12} m \left[n_1^{(\gamma)} (x_2)^2 + n_2^{(\gamma)} (x_1)^2 \right] \end{aligned} \quad [2.98]$$

This Helmholtz energy of the layer depends on the composition of the liquid bulk phase.

The chemical potentials of the components in each of the phases are then:

$$\begin{aligned} \mu_1 &= \left(\frac{\partial G^{(\gamma)}}{\partial n_1}\right)_{T,P,n_1,\sigma} = g_1^{0(\gamma)} + \text{R}T \ln x_1^{(\gamma)} + \text{N}_\text{a} z w_{12} l \left(x_2^{(\gamma)}\right)^2 \\ &+ \text{N}_\text{a} z w_{12} m x_2^2 - \sigma A_M \end{aligned} \quad [2.99]$$

$$\begin{aligned} \mu_2 &= \left(\frac{\partial G^{(\gamma)}}{\partial n_2}\right)_{T,P,n_1,\sigma} = g_2^{0(\gamma)} + \text{R}T \ln x_2^{(\gamma)} + \text{N}_\text{a} z w_{12} l \left(x_1^{(\gamma)}\right)^2 \\ &+ \text{N}_\text{a} z w_{12} m x_2^2 - \sigma A_M \end{aligned} \quad [2.100]$$

We shall now calculate the composition of the surface layer and the surface tension when we know the composition of the liquid and the surface tensions of the two pure liquids.

2.7.3. *Surface tension and composition of the surface layer of a strictly-regular solution*

In conditions of equilibrium, the above chemical potentials are equal to the ordinary chemical potentials of the liquid solution, which, for a strictly-regular solution, are:

$$\mu_1 = g_1^{0(\gamma)} + RT \ln x_1 + N_a z w_{12} x_2^2 \qquad [2.101a]$$

$$\mu_2 = g_2^{0(\gamma)} + RT \ln x_2 + N_a z w_{12} x_1^2 \qquad [2.101b]$$

As the right-hand sides of relations [2.99] and [2.101a] are identical, by using relation [2.54], we find:

$$\sigma A_M = g_1^{0(\gamma)} - g_1^0 + RT \ln \frac{x_1^{(\gamma)}}{x_1} + N_a z w_{12} l \left[\left(x_2^{(\gamma)} \right)^2 - x_2^2 \right] + N_a z w_{12} m x_2^2 \qquad [2.102]$$

Similarly, between the right-hand sides of equations [2.100] and [2.102], the equality gives us:

$$\sigma A_M = g_2^{0(\gamma)} - g_2^0 + RT \ln \frac{x_2^{(\gamma)}}{x_2} + N_a z w_{12} l \left[\left(x_1^{(\gamma)} \right)^2 - x_1^2 \right] + N_a z w_{12} m x_1^2 \qquad [2.103]$$

By applying relation [2.103] to the pure component 1, we obtain:

$$\sigma_1 A_M = g_1^{0(\gamma)} - g_1^0 \qquad [2.104]$$

Similarly, by applying relation [2.102] to the pure component 2, we find:

$$\sigma_2 A_M = g_2^{0(\gamma)} - g_2^0 \qquad [2.105]$$

In light of these equalities, we can write relations [2.102] and [2.103] respectively in the forms:

$$\sigma = \sigma_1 + \frac{RT}{A_M}\ln\frac{x_1^{(\gamma)}}{x_1} + \frac{N_a z w_{12} l}{A_M}\left[\left(x_2^{(\gamma)}\right)^2 - x_2^2\right] + \frac{N_a z w_{12} m}{A_M} x_2^2 \qquad [2.106]$$

$$\sigma = \sigma_2 + \frac{RT}{A_M}\ln\frac{x_2^{(\gamma)}}{x_2} + \frac{N_a z w_{12} l}{A_M}\left[\left(x_1^{(\gamma)}\right)^2 - x_1^2\right] + \frac{N_a z w_{12} m}{A_M} x_1^2 \qquad [2.107]$$

Thus, we obtain two symmetrical relations which were put forward by Schuchovitsky and Guggenheim.

If, in the various relations, we make $w_{12} = 0$, we obtain the same results as those found for perfect solutions in section 2.3.1.

Similarly, by setting $w_{12} = 0$, $x_2^{(\gamma)} = 0$, $x_2 = 0$, and thus $x_1^{(\gamma)} = 1$ and $x_1 = 1$, we find the same result as for the surface energy of the pure substance in relation [2.104].

2.7.4. *Monolayer model and interface tension between two strictly-regular solutions*

The model developed in section 2.7.3 can be used to calculate the surface tension of the interface between two liquid solutions α and β (Figure 2.5) constituting strictly-regular solutions and, in particular, for strictly-regular solutions of the same components but with different compositions and immiscible.

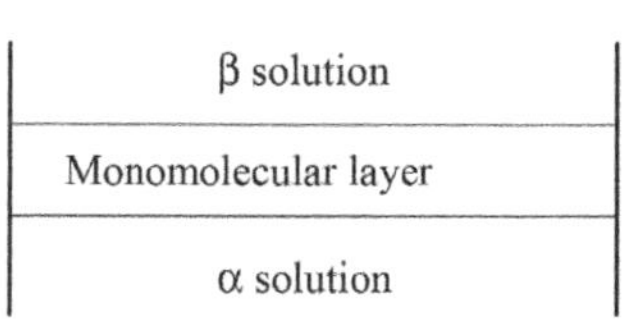

Figure 2.5. *Solutions separated by an interface in the monomolecular layer model of an interface*

The Gibbs energy linked to the monolayer can be calculated as before, and relation [2.85] is replaced by the expression:

$$\begin{aligned} G^{(\gamma)} &= n_1^{(\gamma)}\left[g_1^{0(\gamma)} + RT\ln x_1^{(\gamma)} + N_a z w_{12} l \left(x_2^{(\gamma)}\right)^2 \right] \\ &+ n_2^{(\gamma)}\left[g_2^{0(\gamma)} + RT\ln x_2^{(\gamma)} + N_a z w_{12} l \left(x_2^{(\gamma)}\right)^2 \right] \\ &+ \left(n_1^{(\gamma)} + n_2^{(\gamma)}\right) N_a z w_{12} m \left[\left(x_2^{(\gamma)} - x_2^{(\alpha)}\right)^2 + \left(x_2^{(\gamma)} - x_2^{(\beta)}\right)^2 \right] \end{aligned} \quad [2.108]$$

The chemical potential is equal to the complete chemical potential, and thus it can be calculated on the basis of the generalized Gibbs energy of the surface:

$$\mu_i^{(\alpha)} = \left(\frac{\partial G_\sigma^{(\gamma)}}{\partial n_i^{(\gamma)}} \right) = \left(\frac{\partial G^{(\gamma)}}{\partial n_i^{(\gamma)}} \right) - \sigma A_M \quad [2.109]$$

Hence, instead of expressions [2.99] and [2.100], we find relations [2.110] and [2.111], which are written:

$$\begin{aligned} \mu_1^{(\alpha)} &= g_1^{0(\gamma)} + RT\ln x_1^{(\gamma)} + N_a z w_{12} l \left(x_2^{(\gamma)}\right)^2 \\ &+ N_a z w_{12} m \left[\left(x_2^{(\gamma)} - x_2^{(\alpha)}\right)^2 + \left(x_2^{(\gamma)} - x_2^{(\beta)}\right)^2 + 2x_2^{(\gamma)}\left(2x_2^{(\gamma)} - x_2^{(\alpha)} - x_2^{(\beta)}\right) \right] - \sigma A_M \end{aligned} \quad [2.110]$$

$$\begin{aligned} \mu_2^{(\alpha)} &= g_2^{0(\gamma)} + RT\ln x_2^{(\gamma)} + N_a z w_{12} l \left(x_1^{(\gamma)}\right)^2 \\ &+ N_a z w_{12} m \left[\left(x_2^{(\gamma)} - x_2^{(\alpha)}\right)^2 + \left(x_2^{(\gamma)} - x_2^{(\beta)}\right)^2 + 2x_1^{(\gamma)}\left(2x_2^{(\gamma)} - x_2^{(\alpha)} - x_2^{(\beta)}\right) \right] - \sigma A_M \end{aligned} \quad [2.111]$$

If we now consider that the two α and β phases are two strictly-regular solutions of the same components, then the Gibbs energy terms for the pure components $g_2^{0(\gamma)}$ on the one hand and $g_1^{0(\gamma)}$ on the other are identical in the previous two solutions, and expressions [2.106] and [2.107] are replaced by:

$$\sigma = \frac{RT}{A_M}\ln\frac{x_1^{(\gamma)}}{x_1^{(\alpha)}} + \frac{N_a z w_{12} l}{A_M}\left[\left(x_2^{(\gamma)}\right)^2 - \left(x_2^{(\alpha)}\right)^2\right]$$
$$+\frac{N_a z w_{12} m}{A_M}\left[\left(x_2^{(\gamma)} - x_2^{(\alpha)}\right)^2 + \left(x_2^{(\gamma)} - x_2^{(\beta)}\right)^2 - 2x_1^{(\gamma)}\left(2x_2^{(\gamma)} - x_2^{(\alpha)} - x_2^{(\beta)}\right)\right] \quad [2.112]$$

$$\sigma = \frac{RT}{A_M}\ln\frac{x_2^{(\gamma)}}{x_2^{(\alpha)}} + \frac{N_a z w_{12} l}{A_M}\left[\left(x_1^{(\gamma)}\right)^2 - \left(x_1^{(\alpha)}\right)^2\right]$$
$$+\frac{N_a z w_{12} m}{A_M}\left[\left(x_2^{(\gamma)} - x_2^{(\alpha)}\right)^2 + \left(x_2^{(\gamma)} - x_2^{(\beta)}\right)^2 + 2x_1^{(\gamma)}\left(2x_2^{(\gamma)} - x_2^{(\alpha)} - x_2^{(\beta)}\right)\right] \quad [2.113]$$

Note that these two relations can be used to calculate the unknowns, which are the surface tension of the solution and the composition of the monomolecular layer. We can see that the result is a long way removed from Antonov's law, which we saw in section 2.4, and which seems to be an empirical law.

NOTE.– If we apply relations [2.112] and [2.113] to perfect solutions, we obtain a surface tension of 0, which confirms that two perfect solutions of the same components but different compositions are always miscible and are never separated by an interface.

2.7.5. *Critique of the monomolecular layer model*

The monomolecular model which we have just used is extremely basic, but yields correct results in relation to those obtained directly by experimentation, as shown by Figure 2.6 in the case of ether–acetone solutions. This figure compares the experimental results and those from the previous calculation for the variations in surface tension of the ether–acetone solution as a function of the molar fraction of acetone. The straight line in Figure 2.6 represents the same variation if the solution were perfect.

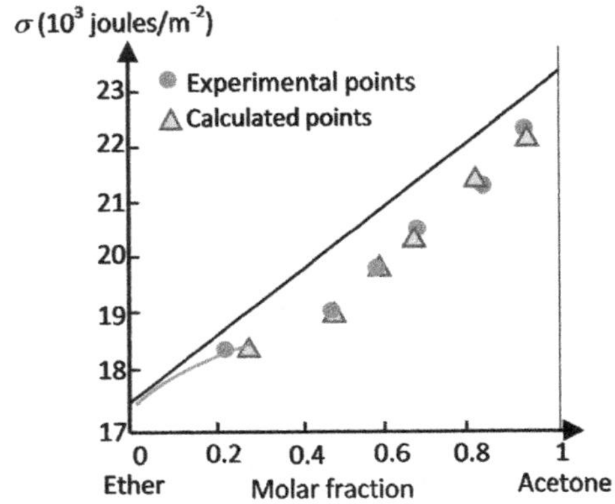

Figure 2.6. *Surface tension of the ether–acetone mixture: comparison between experimental results and the monomolecular layer model of the strictly-regular solution*

However, we shall see that this model is not consistent. In order to do so, we shall examine two problems.

The first deals with the expression of the complete chemical potentials; the second with Gibbs' law of adsorption.

We have seen that at equilibrium, the complete chemical potential of a component of the liquid bulk phase becomes equal to the chemical potential of that component by the zeroing of the lateral chemical potentials (see section 2.2.3). However, if we calculate the complete chemical potential using the last equation [2.15] using expression [2.97], we obtain:

$$\left[\mu_1^{(\alpha)}\right] = g_1^{0(\gamma)} + RT\ln x_1 + \mathrm{N_a} z w_{12} x_2^2 - 2\frac{n_1^{\gamma} + n_2^{(\gamma)}}{n_1 + n_2} m\left(x_2 - x_2^{(\gamma)}\right)x_2 \qquad [2.114]$$

The comparison with expression [2.100] shows that these two relations are differentiated by the final term in [2.114]. Even though the last term may be negligible, as shown by the correct correspondence of the numerical results with the experience, its presence demonstrates the insufficiency of our model.

We can also show that the monomolecular layer model does not satisfy Gibbs' adsorption equation. Indeed, Gibbs' formula [2.46] for a strictly-regular solution is written:

$$\frac{\mathrm{d}\sigma}{\mathrm{d}x_2} = -\frac{\Gamma_{21}}{x_2}\left[\mathrm{R}T + 2\mathrm{N}_{\mathrm{a}}zw_{12}x_1x_2\right] \qquad [2.115]$$

However, by differentiating expressions [2.106] and [2.107], we obtain:

$$\frac{\mathrm{d}\sigma}{\mathrm{d}x_2} = -\frac{\Gamma_{21}}{x_2}\left[\mathrm{R}T + 2\mathrm{N}_{\mathrm{a}}zw_{12}(1-m)x_1x_2\right] \qquad [2.116]$$

Whilst the two formulae [2.115] and [2.116] are indeed identical for a perfect solution, the difference between them shows that the strictly-regular monomolecular layer model is a little simplistic. It has been shown that by introducing two monomolecular layers, the result could be improved.

Our model also supposed that the two molecules of the solution had identical dimensions. Defay and Prigogine (see bibliography) developed a model that is applicable when the dimensions of the molecules are different.

The monomolecular layer model is still very approximate in the case of interfaces between solutions. A better description of the interfaces is obtained by replacing the monomolecular layer with overlapping sheets that are thin enough to consider that, in each of them, the quantities of material are the same at all points, whether or not a sheet is monomolecular. Each sheet is then treated as a non-autonomous phase in contact with two uniform phases of compositions different from its own.

3

Surfaces of Solids and Interfaces

The study of the surface of a solid differs greatly from that of a liquid. This difference stems essentially from the extremely limited mobility of the molecules at the surface of a solid and from a very great difference between the shearing moduli, which is zero in the case of a liquid. Thus, any increase in the area of the surface of a liquid is always attended by an increase in the number of surface atoms without deformation due to the relaxation of the stresses, which is not the case with a solid. Consequently, solid surfaces are usually more rigid and irregular. If the solid is a crystal, toward the outside of the faces, the microcrystals may exhibit vertices or peaks where the atoms present enjoy different environments, making for a very heterogeneous surface.

Another phenomenon distinguishes the surface of a solid from that of liquids. Solid surfaces are almost always tarnished by foreign substances attracted and held to the solid by adsorption (see Chapter 6). Thus, it is extremely difficult to keep the surface of a solid clean for longer than a few moments.

3.1. Surface tension and the surface energy of solids

The equivalence between surface tension and surface energy noted for liquids (see section 1.2.3) no longer applies in the case of solids. The area of the surface of a solid can be modified by varying the number of surface atoms where there is no elastic strain or by applying elastic strain to a constant number of surface atoms.

It is possible to increase the surface area of a solid by creating two surfaces by severing the bonds between molecules or atoms in the bulk phase without altering the relative positions of those molecules or atoms (see Figure 3.1).

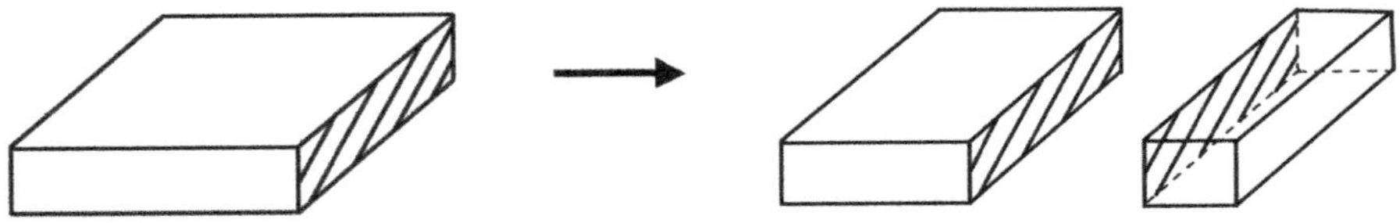

Figure 3.1. *Increase in area, with an increase in the number of surface atoms, by cleaving with no deformation*

The mean area of an atom or molecule does not change, and the total number of atoms or molecules on the surface increases. This is the same situation as for a liquid, as discussed in section 1.2.3, with the same mechanical and thermodynamic equivalence – i.e. equality of the surface tension and the surface energy.

It is also possible for the area to increase due to elastic strain in response to traction exerted on both sides of a solid (Figure 3.2). This traction causes the elongation of the faces in parallel to the direction of traction. Thus, the area increases but the total number of atoms or molecules present at the surface remains unaltered. The mean area of an atom or a molecule decreases and the interface is elastically deformed, meaning that we have:

$$\frac{\partial \sigma}{\partial A} \neq 0 \qquad [3.1]$$

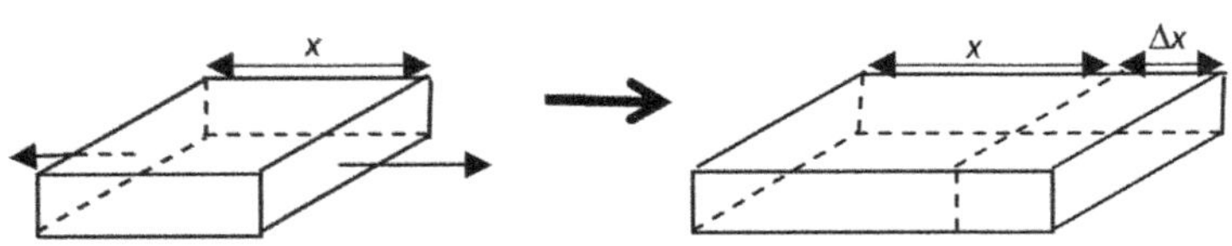

Figure 3.2. *Increasing the surface area of a solid by traction with the same number of surface atoms*

This modifies the calculation of the free energy differential in relation to the area of the surface, at constant temperature, volume and quantity of material in relation [1.53], which is supposed to be valid for deformed surfaces. It now leads to an isotropic interface stress σ^* given by:

$$\sigma^* = \left(\frac{\partial F^{(\gamma)}}{\partial A}\right)_{T,V,N} = \sigma + \left(A\frac{\partial \sigma}{\partial A}\right)_{T,V,N} = \sigma + \left(\frac{\partial \sigma}{\partial \ln A}\right)_{T,V,N} \quad [3.2]$$

This stress is also known as the *surface tension*. The elastic deformation (strain) ε of the surface is such that:

$$\mathrm{d}(\ln A) = \frac{\mathrm{d}\sigma}{A} \cong \frac{\delta A}{A} = \varepsilon \quad [3.3]$$

Hence, we find the expression of the surface tension:

$$\sigma^* = \sigma + \left(\frac{\partial \sigma}{\partial \varepsilon}\right)_{T,V,N} \quad [3.4]$$

This relation was devised by Shuttleworth.

Thus, depending on whether or not the surface is deformed, we use either relation [3.4] or [1.42] to obtain the surface tension.

3.2. Surface energy of a pure crystallized solid: the macroscopic approach

For an initial, crude model, we shall consider the flat surface of a monatomic crystal, and express the surface energy, taking account only of the pair energies ε_{AA} (a value which is negative by definition) with the nearest neighbors of an atom or a molecule. The surface of the crystal is in the presence of its vapor, but the distances between species (atoms or molecules) in the vaporous state are such that the interactions between surface atoms or molecules and those in the vapor are null, as are those between the species making up the vapor. The crystalline arrangement of the surface thus constitutes a monomolecular layer containing N molecules.

In the plane of the layer, *zl* denotes the number of nearest neighbors of a molecule and *zm* the number of pairs between a molecule in the surface layer and one in the adjacent layer; z is the coordination number of the crystal. We therefore have $2m+l=1$.

We form a surface by cleaving a crystal in parallel to a dense plane (as in the scenario where relation [1.42] is applied). Such cleaving creates two fresh surfaces, each containing N molecules. The number of AA pairs broken is zmN, but the number of near neighbors in each layer is not altered by the cleaving. Thus, the increase in energy created by that cleaving is:

$$e^{xs} = -zmN\varepsilon_{\mathrm{AA}} \qquad [3.5]$$

If we consider a molar layer, it contains $\mathrm{N_a}$ molecules ($\mathrm{N_a}$ is Avogadro's number) and its area is A_M. Thus, the surface energy is:

$$\sigma = -\frac{zm\mathrm{N_a}}{2A_M}\varepsilon_{\mathrm{AA}} \qquad [3.6]$$

The pair energy is linked to the enthalpy of sublimation ($\mathrm{sol} \to \mathrm{vap}$ transformation) by the relation:

$$\varepsilon_{\mathrm{AA}} = \frac{2\Delta_{\mathrm{sol}\to\mathrm{vap}}H}{z\mathrm{N_a}} \qquad [3.7]$$

If we choose the (1,1,1) face of a cubic crystal with centered faces, then $z = 12$ and $m = ¼$. By feeding those values back into the previous relations, we find:

$$\sigma = \frac{m\Delta_{\mathrm{sol}\to\mathrm{vap}}H}{A_M} = \frac{\Delta_{\mathrm{sol}\to\mathrm{vap}}H}{4A_M} \qquad [3.8]$$

Comparison with experimental data shows that, in fact, the values deduced by this model are much higher than the true values, because the hypotheses made are very crude: an identical degree of coordination in the plane before and after cleaving, homogeneous surface (i.e. a crystal with no corners or vertex atoms), etc.

3.3. Surface energy in a mesoscopic model

In a mesoscopic model, we differentiate the various positions of the atoms at the surface of a solid. Envisage a solid in the presence of its own vapor. The reasoning applied in section 1.2.4 remains largely unchanged if we apply it to the construction of the crystal from its vapor, although there are two major differences:

– The chemical bond between atoms inside a bulk is sensitive to the size of that bulk. This effect is linked to the quantum nature of the chemical bond and to the origin of the energy bands in solids: it is the large number of atoms (and therefore of atomic orbitals) which creates a large number of molecular levels. Those molecular levels are situated in a finite energy space, and are therefore very close together, and under the influence of thermal expansion they meld into one. In a bulk containing a small number of atoms, a certain quantification can therefore substitute the band structure. This phenomenon is more marked when the electrons are shared over a long distance. It should occur less strongly in molecular or highly-ionic crystals than in metals. Thus, it may be necessary to take account of an effect of size on the chemical bond even within the bulk, whose enthalpy would be: $\Delta_{qu}(H)$.

– Unlike in the case of a liquid, mechanical equilibrium is not achieved in the solid, and its form will not necessarily be spherical. Thus, we must distinguish between various types of structural elements at the surface, depending on whether those elements are situated on the corners, edges or faces, which may, themselves, differ from one to another.

The construction of the crystal from its vapor, therefore, will involve the following transformations:

– the passage of a molecule from the vapor to the bulk of the solid: vap → solid;

– the passage of a molecule from the vapor to the vertices of the surface: vap → vertices;

– the passage of a molecule from the vapor to the edges: vap → edges;

– the passage of a molecule from the vapor to a face of the crystal: vap → face.

Instead of relation [1.50], we obtain:

$$\begin{aligned}\sigma A_M &= \Delta_{\text{qu}}(H) + x_{\text{vertices}}\left[\Delta_{\text{vap}\to\text{s}}(H) - \Delta_{\text{vap}\to\text{vertices}}(H)\right] \\ &+ x_{\text{edges}}\left[\Delta_{\text{vap}\to\text{solid}}(H) - \Delta_{\text{vap}\to\text{edges}}(H)\right] \\ &+ \sum x_{\text{face}}\left[\Delta_{\text{vap}\to\text{solid}}(H) - \Delta_{\text{vap}\to\text{face}}(H)\right]\end{aligned} \quad [3.9]$$

As in the previous case, σ is a positive value.

We cannot eliminate the term σ in equation [3.9]. Indeed:

– the term $\Delta_{\text{qu}}(H)$ does not necessarily obey a law of proportionality with the inverse of the radius of the particle;

– the number of vertices is independent of the size, because it is determined only by the shape;

– the size of edges varies, in a given shape, in a linear fashion with particle size.

Thus, the interfacial energy will generally depend on the shape and size of the particles.

It has been shown that certain crystalline surfaces created by cleaving along the crystallographic planes of solids exhibit interfacial energies tending toward infinity: such is the case if there is a dipolar moment perpendicular to the surface plane. In this case, it is possible that the face will not appear in the true solid, or that its components will adopt different positions those to predicted by a fixed lattice cross-section: this phenomenon is known as surface reconstruction, and tends to decrease the interfacial energy.

The reasoning we employed for the interface between the solid and its vapor would also apply to the interface between the solid and the same substance in the molten state.

3.4. Effective surface energy: the Wulff crystal

We saw in Chapter 1 that in the case of the formation of a liquid, it was possible to define a unique interfacial energy. In addition, the equilibrium of the mechanical forces means that the shape must be spherical. Equation [3.9], which corresponds to the excess Gibbs energy in the case of a solid, is substantially more complicated. However, we shall now define a method whereby it can be applied directly to the case of solids, which leads us to define the *effective interfacial energy*.

Consider bulks where each dimension is large on the chemical bond scale. Therefore, we can discount:

– the term $\Delta_{\text{qu}}(H)$;

– the term due to the vertices, whose number remains constant in a given polyhedron;

– the term due to the edges (which increase in number with the size) in favor of the term due to the faces (which increase in number with the square of the size).

Relation [3.9] then becomes:

$$\sigma.A_M = \sum_{\text{faces}} x_{\text{faces}} \left[\Delta_{\text{vap}\to\text{solid}}(H) - \Delta_{\text{vap}\to\text{face}}(H) \right] \qquad [3.10]$$

We can therefore write:

$$\sigma A_M = \sum_{i \text{ faces}} \sigma_i A_{M_i} \qquad [3.11]$$

where A_{M_i} and σ_i respectively represent the area and the interfacial energy of the face i.

We shall now show that in a solid of any given polyhedral shape, the distribution of the interfacial energies leads to the existence of the shape with the least energy, independent of size, using the approximation that the interfacial energy is, itself, independent of the size.

Thus, consider a solid of any polyhedral form with constant volume V. Pick any given point P inside the solid, and let h_i represent the distance from that point to the plane containing the face i (see Figure 3.3).

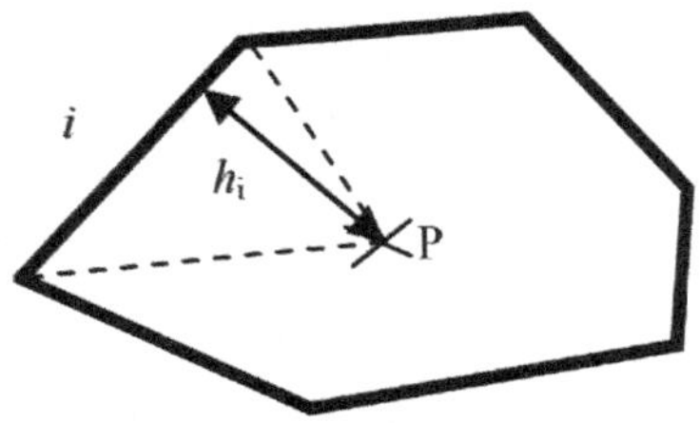

Figure 3.3. *Finding the shape of the crystal with the least energy*

The volume of the elementary solid, which is determined by point P and the vertices of the face i, is:

$$V_i = \int_0^{h_i} A_{M_i} \left(\frac{u}{h_i}\right)^2 \mathrm{d}u = \frac{h_i A_{M_i}}{3} \qquad [3.12]$$

The total volume of the solid, therefore, is:

$$V = \frac{1}{3}\sum_{i=1}^{i_m} A_{M_i} h_i \qquad [3.13]$$

where i_m is the number of faces of the polyhedron.

The total interfacial energy is:

$$W = \sum_{i=1}^{i_m} \sigma_i A_{M_i} \qquad [3.14]$$

The shape with the least energy is such that for any infinitesimal transformation at constant volume, the differential of the total interfacial energy is zero. Gibbs gave us the solution to this problem. For each face i, the ratio σ_i / h_i is a real constant k, independent of i. Thus, we have $W = 3kV$, and for any transformation $\mathrm{d}W = 3k\mathrm{d}V$, so $\mathrm{d}W$ is zero at constant volume.

It is helpful to take the center of inertia of the solid as point P. We can see that the faces whose interfacial energy is high are situated a long way from the center of the solid: if their plane of equilibrium is further away than the plane containing the intersection of the adjacent faces, they will not appear in the solid. Thus, the number i_m is, itself, set by the relative σ_i values. Hence, the shape of a crystal of finite size depends not on its size, but solely on the different energies of its faces: crystals of differing sizes can be deduced from one another by homothety, and the fraction of the area due to each face will not depend on the size, so no matter what the value of i between 1 and i_m, the ratio $\frac{\sigma_i}{\sigma} = \alpha_i$ is independent of the size. The crystal thus described is known as the *Wulff crystal.*

It is therefore possible to characterize the Wulff crystal using a single length a, chosen in any given (but fixed) crystallographic direction, and from the geometry for each case we can deduce a volumic form factor, hereafter denoted as Ψ_v, and a surface form factor, written as Ψ_s, independent of a and such that, for the volume and the surface of the crystal, we have:

$$V = \Psi_v.a^3 \qquad [3.15]$$

and

$$A_{\mathrm{M}} = \Psi_s.a^2 \qquad [3.16]$$

This means that we can write the total interfacial energy as:

$$W = a^2.\Psi_s.\sum_{i=1}^{i_m}\sigma_i\alpha_i \qquad [3.17]$$

By introducing the radius r of the sphere of the same volume, i.e. such that $\frac{4\pi}{3}r^3 = \Psi_v.a^3$, we find:

$$W = 4\pi r^2\left\{\Psi_s\left(3\Psi_v\right)^{-2/3}\left(4\pi\right)^{-1/3}\sum_{i=1}^{i_m}\sigma_i\alpha_i\right\} \qquad [3.18]$$

The term in chain brackets here has the dimension of an interfacial energy, and does not depend on the size of the crystal, but on the different interfacial energies, which:

– directly define the terms σ_i;

– indirectly, by way of the morphology, define the terms α_i ,Ψ_v and Ψ_s.

W is the *effective interfacial energy of the crystal.*

3.5. Interfacial energy between two solids

Up until now, our discussion has been based on the assumption that all the solids were perfect and therefore had a molar enthalpy of formation which depended only on the temperature. This approximation is justified in the study of the condensation of a gas (or a liquid), because the variations in the chemical potential as a function of the composition are much greater in the fluid phase and, in practice, render the same effects in the solid negligible. This is no longer true, though, with contact between two solids, because it is the impurities in the solid which are the species that determine the chemical potentials.

We can apply the reasoning process illustrated in section 3.3 to the appearance of one solid from another solid, provided we use the quasi-chemical formalism of the structural elements:

– the fluid is replaced by the original solid phase;

– the molecules or ions are replaced by the structural elements.

Let s1 and s2 represent the two solids in contact. We are led to the same formalism as in section 3.3:

$$\begin{aligned}\sigma A_M &= \Delta_{\text{qu}}(H) + x_{\text{vertices}}\left[\Delta_{\text{s1->s2}}(H) - \Delta_{\text{s1->s2vertices}}(H)\right]\\ &+ x_{\text{edges}}\left[\Delta_{\text{s1->s2}}(H) - \Delta_{\text{s1->s2edges}}(H)\right]\\ &+\sum_i x_i\left[\Delta_{\text{s1->s2}}(H) - \Delta_{\text{s1->s2face}}(H)\right]\end{aligned} \qquad [3.19]$$

In the case of a solid–fluid interface, we adopted the hypothesis that the enthalpy of condensation must lie between 0 and the value obtained in the bulk of the solid. We shall see, though, that this point needs to be reconsidered in the case of a solid–solid interface, integrating the concept of compatibility of the crystalline lattices and of the molar volumes.

The three-dimensional order of crystallized solids requires a great deal of entropy. The reason for it is to maximize the interactions (van der Waals forces, hydrogen bonds, electrostatic forces, delocalization of electrons in a metal) between the structural elements, in view of the geometric stresses stemming from the shape of the molecules or the size of the ions. The whole of the solid, therefore, creates a periodic potential – particularly in the space surrounding a structural element, which therefore, in the perfect solid at 0K, occupies the position of least energy.

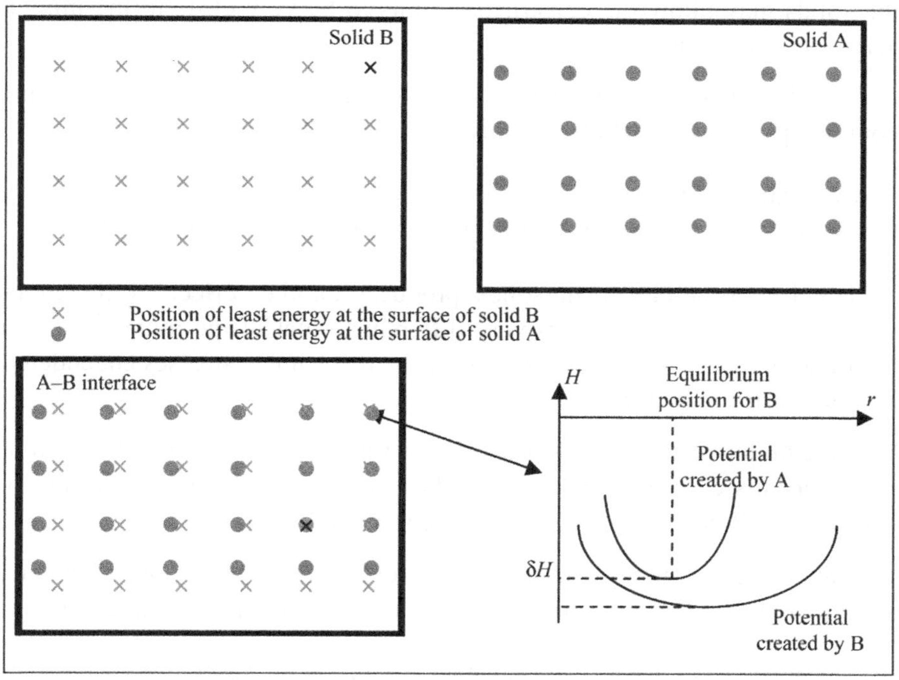

Figure 3.4. *The importance of compatibility between the crystalline lattices*

As an initial approximation (notably if we overlook the phenomenon of surface reconstruction), a structural element situated at the surface is subject to a potential which is half of that at the core. Let us now consider the situation at a flat interface between two solids, A and B: take a structural element situated both in the minima of potentials created by A and B: its molar enthalpy is half the sum of the enthalpies in the two solids, and we see the same situation as with the solid–fluid case, if that reasoning process were not usually limited to a small fraction of the construction units situated at the interface. Indeed, the potential of the solids has the same period as their spatial arrangement, so, in the general case of solids whose crystallographic characteristics are different, it is not possible for neighboring construction units to be situated at the minima of potential both for A and B (see Figure 3.4).

The least unfavorable scenarios are those where:

– either the lattices are fairly similar: in this case, the small size of one of the solids may allow the construction units to be very close to the minima across the whole surface;

– or the spatial periods of the lattices are commensurate, which ensures common periodicity that affects only a fraction of the sites.

Generally speaking, we see the formation of a grain joint, where the reconstruction of the two interfaces gives rise to an amorphous layer.

The lack of mobility in solids produces another effect: if the molar volume of the solid formed is very different to that of the initial solid, the surrounding phase cannot accommodate the mechanical stresses engendered. Two scenarios may occur:

– the molar volume of the phase formed is smaller than that of the initial phase: in this case, we see the formation of interfaces between A and the surrounding fluid and between B and the surrounding fluid;

– the molar volume is larger: then we see the creation of mechanical stresses and the appearance of an additional term in the interfacial energy.

The interfacial energy between two solids, therefore, may be extremely low if they have different compositions but their crystalline lattices are similar in terms of symmetry and mesh parameters. We shall see that this results in the phenomena of heterogeneous primary nucleation from the fluid phases, which ultimately results in epitaxy. On the other hand, that interfacial energy may be very high if the lattices are very different in terms of their molar volumes and from a crystallographic point of view.

3.6. Interfaces between pure solids and liquids

We shall now touch on a certain number of phenomena linked to the existence of an interface between a pure solid and a pure liquid – phenomena which are connected to the properties of wetting and adhesion of the liquid to the solid, and to numerous applications such as detergence.

3.6.1. *Spreading and angle of contact of a liquid on a solid*

As we did for the drop of liquid on a liquid (see section 2.6), we can study the behavior of a drop of liquid placed on a solid.

Consider a flat solid surface upon which a drop of a liquid is deposited. That drop may or may not spread. Suppose it does not spread and that the liquid does not react with the solid. Let θ be the angle of contact formed by the tangent to the drop at the "triple" point where the three phases meet: the solid, the liquid and the surrounding gas (see Figure 3.5(a)). This angle may have any value at all between 0 and π radians, and is sometimes referred to as the *link angle*. The three surface tensions σ_s, σ_L and σ_{sL} are represented by three vectors applied at that triple point, tangential to the path of the respective interfaces.

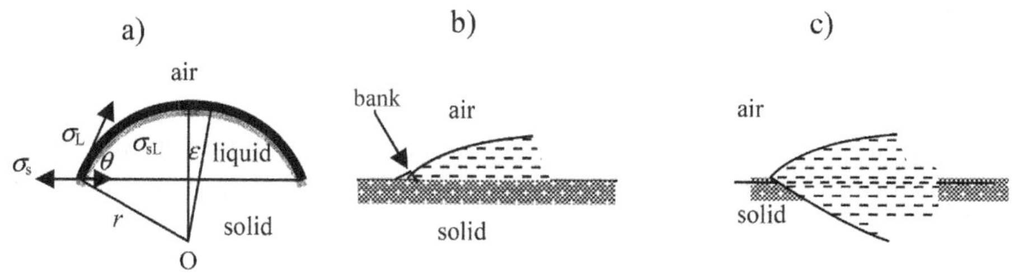

Figure 3.5. *Lie of the drop on a solid: a) metastable equilibrium; b) equilibrium reached at the triple point; c) all equilibria are achieved*

The condition of mechanical equilibrium, therefore, is written:

$$\vec{\sigma}_s + \vec{\sigma}_L + \vec{\sigma}_{sL} = 0 \quad [3.20]$$

However, the representation of the interfacial tension vectors in Figure 3.5(a) does not satisfy this condition. Experience tells us that within a short period of time, a state of mechanical equilibrium of the drop, called *temporary equilibrium*, is reached, and Young and Dupré express that state by projecting the vectorial equality onto the plane of the substrate, which gives us:

$$\sigma_s = \sigma_L \cos\theta + \sigma_{sL} \qquad [3.21]$$

Let us show that this relation corresponds to a state of equilibrium.

The generalized capillary Gibbs energy for the system represented in Figure 3.5(a) would be written as:

$$G_\sigma = G(T,P) + \sigma_L A_L + \sigma_s A_s + \sigma_{sL} A_{sL} \qquad [3.22]$$

Let us now evaluate each of the interfacial areas:

– the area of the interface between the solid and the liquid is that of a disk whose radius is $r \sin\theta$:

$$A_{sL} = \pi r^2 \sin^2\theta \qquad [3.23]$$

– the area of the gas–solid interface is given by the initial area of the solid in the absence of the drop A_0, less the area of the previous liquid–solid interface, which is:

$$A_{Gs} = A_0 - \pi r^2 \sin^2\theta \qquad [3.24]$$

– the area of the interface between the gas and the liquid is that of the spherical cap with radius r, centered at O and observed from that point at the solid angle Ω:

$$A_{GL} = \Omega r^2 = 2\pi r^2 (1 - \cos\theta) \qquad [3.25]$$

By introducing the three values of the areas into relation [3.22], we obtain:

$$G_\sigma = G(P,T) + \pi r^2 \sigma_{sL} \sin^2\theta + \sigma_s \left(A_0 - \pi r^2 \sin^2\theta\right) + 2\pi r^2 \sigma_L (1 - \cos\theta) \qquad [3.26]$$

Thus, the capillary Gibbs energy function is a function of the variables T, P, r and θ. Its differential at constant temperature and external pressure, therefore, is of the form:

$$(\mathrm{d}G_\sigma)_{T,P} = \frac{\partial G_\sigma}{\partial r}\mathrm{d}r + \frac{\partial G_\sigma}{\partial \theta}\mathrm{d}\theta = 0 \qquad [3.27]$$

This differential is zero if the system is at equilibrium (be it stable or unstable). When we differentiate expression [3.26], the equilibrium condition takes the form:

$$\begin{aligned}&\left[4\pi r\sigma_L(1-\cos\theta) + 2\pi r\left(\sigma_{sL} - \sigma_s\sin^2\theta\right)\right]\mathrm{d}r \\ &+\left[2\sigma_L\pi r^2\sin\theta + 2(\sigma_{sL} - \sigma_s)\pi r^2\sin\theta\cos\theta\right]\mathrm{d}\theta = 0\end{aligned} \qquad [3.28]$$

The volume of the liquid is that of the spherical sector with radius r, centered at O. Its value is:

$$V_L = \int_0^\theta\left(\pi r^2\sin^2\varepsilon\right)\left(rd\varepsilon\sin\varepsilon\right) = \frac{\pi r^3}{3}\left(2 - 3\cos\theta + \cos^3\theta\right) \qquad [3.29]$$

At equilibrium in the drop, that volume is constant, and therefore its differential is zero:

$$\mathrm{d}V_L = \frac{\partial V_L}{\partial r}\mathrm{d}r + \frac{\partial V_L}{\partial \theta}\mathrm{d}\theta = 0 \qquad [3.30]$$

If we differentiate relation [3.29] and make the two differentials $\mathrm{d}V_\mathrm{L}$ equal, this gives us:

$$\mathrm{d}r = -\frac{r\sin\theta(1+\cos\theta)}{2-\cos\theta+\cos^2\theta}\mathrm{d}\theta \qquad [3.31]$$

By feeding back equation [3.31] into relation [3.28], we obtain the Young–Dupré equation, given by expression [3.21]. Nothing in our calculation specifies whether this is a stable or unstable equilibrium.

The local stable equilibrium at the triple line can be obtained by local deformation of the solid, which then presents a bank with a slight indentation in the solid on both sides (see Figure 3.5(b)). This morphological alteration of the solid is obtained by interfacial and surface diffusion, which explains the long periods of time needed – particularly at low temperature. This modification causes the system to tend toward a state of stable local equilibrium, which is obtained (Figure 3.5(c)) when the curvature of the liquid–solid interface is such that relation [3.20] is satisfied.

For certain liquid and solid couples, the drop cannot achieve equilibrium within a short space of time, and spreads completely over the solid. Thus, the surface tensions are such that:

$$\sigma_s > \sigma_{sL} > \sigma_L \qquad [3.32]$$

We then say that the liquid wets the solid perfectly.

Generally, we obtain the difference:

$$h_t = \sigma_s - (\sigma_{sL} + \sigma_L) = \sigma_L(\cos\theta - 1) \qquad [3.33]$$

This difference is the *coefficient of spreading* (see relation [1.81]) of the liquid over the solid. Equilibrium between the drop and the solid, therefore, can only come about if that spreading coefficient is negative – i.e. if the liquid–solid couple satisfies the dual condition:

$$-2\sigma_L \le h_t \le 0 . \qquad [3.34]$$

3.6.2. *Work of adhesion between a liquid and a solid*

For the work of adhesion between a liquid and a solid, the definition given in section 2.5 for two liquids remains valid, and by applying Dupré's relation [2.75], in light of relation [3.21], we obtain:

$$W_{sL} = \sigma_L(1 + \cos\theta) \qquad [3.35]$$

Thus, the work of adhesion of a solid and a liquid can be calculated if we know the surface tension of the liquid and the contact angle between a drop of the liquid and the solid – two values which it is possible to measure.

3.6.3. *Solid surface in contact with two liquids: displacement of one liquid by another*

Let us now look at what happens when two liquids are brought into contact on the same solid surface. Two scenarios may arise:

– the two liquids may coexist, side by side, on the surface of the solid;

– or else, one of the liquids repels the other, and tends to occupy the whole of the space.

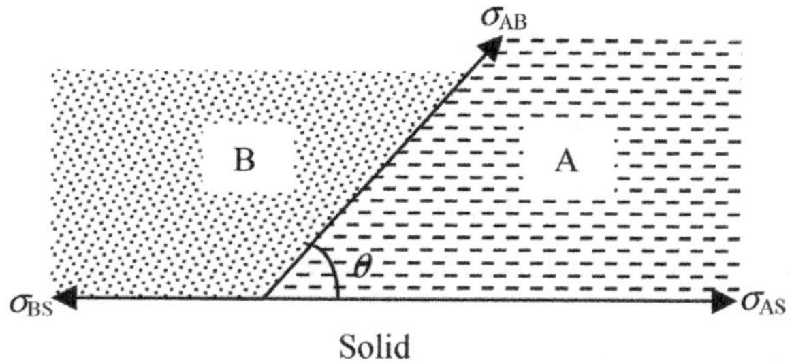

Figure 3.6. *Two liquids coexisting on the surface of a flat solid*

Consider the first case (Figure 3.6). The two liquids A and B may coexist on the flat surface of the solid. Let θ be the angle of contact, which is supposed to be acute on the side of A. The temporary equilibrium condition for this system is:

$$\sigma_{\text{Bs}} = \sigma_{\text{As}} + \sigma_{\text{AB}} \cos\theta \qquad [3.36]$$

We can calculate the interfacial tension between the two liquids σ_{AB}, but we do not know the solid–liquid interfacial tensions σ_{As} and σ_{Bs}. In order to obtain them, consider the work of adhesion of each of the two liquids to the solid. Given relation [2.75], we have:

$$W_{\text{As}} = \sigma_{\text{A}} + \sigma_{\text{s}} - \sigma_{\text{As}} \qquad [3.37a]$$

$$W_{\text{Bs}} = \sigma_{\text{B}} + \sigma_{\text{s}} - \sigma_{\text{Bs}} \qquad [3.37b]$$

By finding the difference, we obtain:

$$W_{\mathrm{As}} - W_{\mathrm{Bs}} = \sigma_{\mathrm{A}} - \sigma_{\mathrm{B}} + \sigma_{\mathrm{Bs}} - \sigma_{\mathrm{As}} \qquad [3.38]$$

In light of relation [3.36], we find:

$$W_{\mathrm{As}} - W_{\mathrm{Bs}} = \sigma_{\mathrm{A}} - \sigma_{\mathrm{B}} + \sigma_{\mathrm{AB}} \cos\theta \qquad [3.39]$$

From this, we deduce:

$$\cos\theta = \frac{\left(W_{\mathrm{As}} - \sigma_{\mathrm{A}}\right) - \left(W_{\mathrm{Bs}} - \sigma_{\mathrm{B}}\right)}{\sigma_{\mathrm{AB}}} \qquad [3.40]$$

If θ_{A} and θ_{B} are the angles of contact of liquids A and B, deposited separately onto the solid surface, then in view of relation [3.35], we have:

$$W_{\mathrm{As}} = \sigma_{\mathrm{A}}\left(1 + \cos\theta_{\mathrm{A}}\right) \qquad [3.41]$$

$$W_{\mathrm{Bs}} = \sigma_{\mathrm{B}}\left(1 + \cos\theta_{\mathrm{B}}\right) \qquad [3.42]$$

These relations, when fed back into expression [3.40], give us:

$$\cos\theta = \frac{\sigma_{\mathrm{A}} \cos\theta_{\mathrm{A}} - \sigma_{\mathrm{B}} \cos\theta_{\mathrm{B}}}{\sigma_{\mathrm{AB}}} \qquad [3.43]$$

We have supposed that the angle θ was acute on the side of A, so $\cos\theta$ must be positive and less than 1, which means we can write the equilibrium condition in the form:

$$\cos\theta = \frac{\sigma_{\mathrm{A}} \cos\theta_{\mathrm{A}} - \sigma_{\mathrm{B}} \cos\theta_{\mathrm{B}}}{\sigma_{\mathrm{AB}}} \qquad [3.44]$$

Freundlich spoke of the *tension of adhesion* (not to be confused with the work of adhesion) of each of the liquids, representing them by the terms $\sigma_{\mathrm{A}} \cos\theta_{\mathrm{A}}$ and $\sigma_{\mathrm{B}} \cos\theta_{\mathrm{B}}$:

$$\tau_{\mathrm{A}} = \sigma_{\mathrm{A}} \cos\theta_{\mathrm{A}} \qquad [3.45a]$$

$$\tau_B = \sigma_B \cos\theta_B \qquad [3.45b]$$

Hence, the equilibrium condition is written:

$$\tau_A - \tau_B < \sigma_{AB} \qquad [3.46]$$

The difference between the tensions of adhesion of the two liquids must be less than their interfacial tension.

In the opposite case, when condition [3.46] is not respected there is no possible equilibrium, and liquid A (which has the higher tension of adhesion) spreads and forces liquid B out.

This phenomenon of displacement is exploited in the assisted extraction of petroleum from porous rocks, but also in detergence, where a detersive substance is added to water to help it to displace grease on the surface of a cloth or a dish.

3.6.4. *Conditions of stability of solid particles at fluid interfaces*

It is easy to see that small particles of numerous solids may be at apparently-stable equilibrium at liquid–fluid interfaces. Such is the case, for example, with a metal needle which, though it is far denser than water, floats on the surface of water if coated with wax. We shall now examine this phenomenon of equilibrium of a small solid at the interface between two fluids.

Consider a solid particle that is small enough for the influence of the surface forces upon it to be stronger than that of gravity. For simplicity's sake, let us say that this solid particle is a rectangular parallelepiped. Place that particle at the interface between two fluids – e.g. a liquid L and air (Figure 3.7).

We can see that at a triple point, the temporary equilibrium condition is written:

$$\sigma_s = \sigma_L \cos\theta + \sigma_{sL} \qquad [3.47]$$

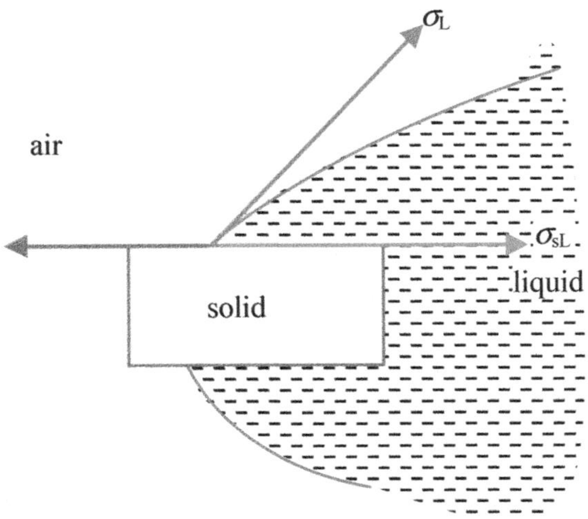

Figure 3.7. *Stability of a solid particle at liquid interfaces*

This relation is absolutely identical to relation [3.21], which gives the condition of non-wetting of the solid surface by the liquid. If this condition is not met, the liquid spreads over the whole of the solid surface, which is then completely surrounded by liquid and therefore cannot remain at the surface. Thus, the condition for stability of the particle is that the liquid must not wet the solid.

Looking now at the case of the interface between two liquids, A and B (Figure 3.8), the equilibrium condition here is:

$$\sigma_{sB} = \sigma_{AB} \cos\theta + \sigma_{sA} \qquad [3.48]$$

Then we again see condition [3.36] for the coexistence of two liquid surfaces on the surface of a solid. If this condition is not fulfilled, the solid particles are rejected from the liquid whose adhesion tension is higher.

The separation of ores by flotation is one of the essential applications of this phenomenon of stability or rejection of solid particles at the interface between two fluids.

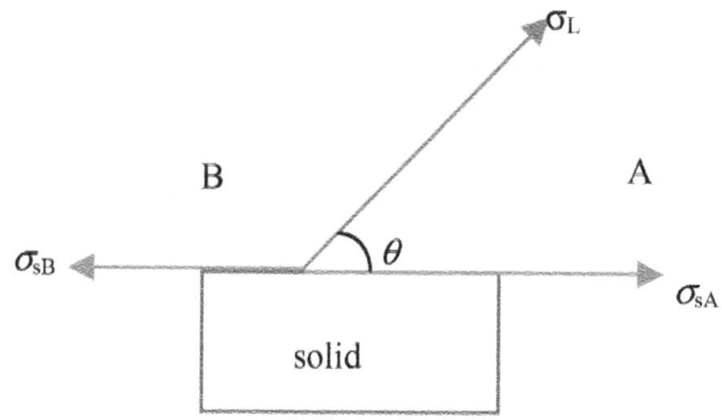

Figure 3.8. *Stability of a solid particle at the interface between two liquids*

3.7. Adsorption of elements of a liquid solution by a solid

We shall now consider the interface between a pure solid and a liquid binary solution (β phase) made up of the two components, 1 and 2. The composition of the liquid in the vicinity of the solid expresses the phenomenon of adsorption of the elements in the liquid solution by the solid. The monolayer model (γ phase), which we used in section 2.7, coupled with the strictly-regular solution model, taking account only of the pair energies, yields the following expression, which was put forward by Desré:

$$\begin{aligned}(\sigma_{sL}-\sigma_s)A_M &= \left[x_1^{(\gamma)}(\sigma_1-W_1)+x_2^{(\gamma)}(\sigma_2-W_2)\right]A_M \\ &+RT\left[x_1^{(\gamma)}\ln\frac{x_1^{(\gamma)}}{x_1}\right]+\left[x_2^{(\gamma)}\ln\frac{x_2^{(\gamma)}}{x_2}\right]+(l+m)\mathrm{N_a}w_{12}x_A x_B \\ &+l\mathrm{N_a}w_{12}x_A^{(\gamma)}-(l+m)\mathrm{N_a}w_{12}\left(x_1^{(\gamma)}x_2+x_2^{(\gamma)}x_1\right)\end{aligned} \quad [3.49]$$

Remember that in this relation, $x_1^{(\gamma)}$ and x_1 denote the molar fraction of component 1 in the monolayer and in the bulk of the liquid, respectively. A_M is the molar area, σ_s and σ_L are the surface tensions of the solid and liquid, each in the presence of its own vapor, σ_1 and σ_2 are respectively the surface tensions of pure liquids 1 and 2 in the presence of their vapor, W_1 is the work of adhesion of the pure liquid 1 to the solid s, $\mathrm{N_a}$ is Avogadro's number and

w_{12} is the parameter representing the difference in the pair energies which characterizes the strictly-regular solution.

The minimization of the free energy in relation to the quantity of component 1 in the monolayer, for a given composition of the bulk solution, gives us:

$$\mathrm{R}T \ln \frac{x_2^{(\gamma)}\left(1-x_2\right)}{x_2\left(1-x_2^{(\gamma)}\right)} = \left(\sigma_1-\sigma_2\right)A_M-\left(W_1-W_2\right)A_M \qquad [3.50]$$
$$-l\mathrm{N_a}w_{12}\left(1-2x_2^{(\gamma)}\right)+\left(l+m\right)\mathrm{N_a}w_{12}\left(1-2x_2\right)$$

Expressions [3.49] and [3.50] can be used to obtain the values of the two unknowns: $x_2^{(\gamma)}$, the composition of the solution in the vicinity of the surface of the solid; and σ_{sL}, the liquid–solid interface energy, for a given composition x_2 of the solution.

Relation [3.49], applied to a dilute solution of component 2 in solvent 1 – i.e. when x_2 tends toward zero – enables us to calculate the variation in the interfacial energy with changing composition. We find:

$$\left(\frac{d\sigma_{sL}}{dx_2^{(\gamma)}}\right)_{x_2\to 0} = \frac{\mathrm{R}T}{A_M}\left[1-\exp\left(-\frac{E_2}{\mathrm{R}T}\right)\right] \qquad [3.51]$$

where E_2 is the energy of adsorption of 2 by the solid at infinite dilution, defined by:

$$E_2=\left(\sigma_2-\sigma_1\right)A_M-\left(W_2-W_1\right)A_M-m\mathrm{N_a}w_{12} \qquad [3.52]$$

The application of relation [3.50] for the dilute solution gives us:

$$\left(\frac{x_2^{(\gamma)}}{x_2}\right)_{x_2^{(\gamma)}\to 0} = \exp\left(-\frac{E_2}{\mathrm{R}T}\right) \qquad [3.53]$$

This expression of the adsorption isotherm is similar to Henry's law.

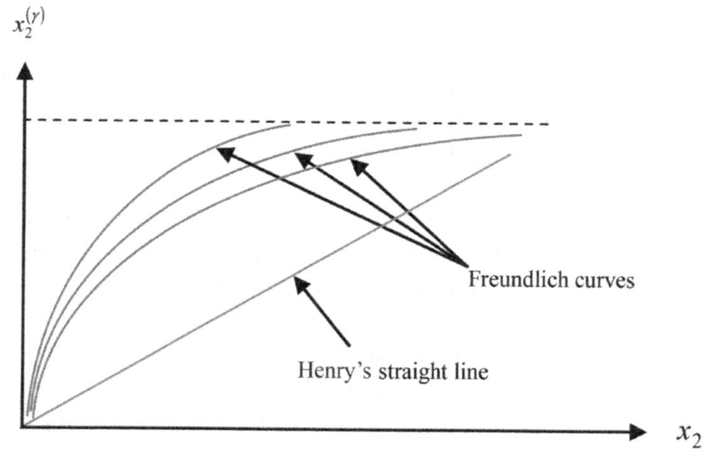

Figure 3.9. *Isotherm of adsorption of a solute to a solid*

We know that such an equation cannot account for all the forms of isotherms of adsorption. The application of equation [3.50] for perfect solutions ($w_{12} = 0$) gives us the homographic form of the isotherm:

$$\left(\frac{x_2^{(\gamma)}}{1-x_2^{(\gamma)}}\right) = \frac{x_2}{1-x_2}\exp\left(-\frac{E_2}{RT}\right) = \frac{Kx_2}{1-x_2} \quad [3.54a]$$

or

$$x_2^{(\gamma)} = \frac{Kx_2}{1+(K-1)x_2} \quad [3.54b]$$

More complex laws such as Freundlich's, which expresses an exponential variation of the quantity of fixed component 2 as a function of the concentration of the solution (Figure 3.9), have been put forward, but no attractive models have been developed to illustrate them.

3.8. Electrocapillary phenomena

Electrocapillarity is the combination of electrical phenomena and surface phenomena. It is particularly useful when studying the double electrochemical layers found at electrodes.

3.8.1. *Definition of electrocapillarity*

The phenomenon of electrocapillarity consists of a variation in the surface tension of an α phase in contact with another β phase when the potential difference between the two phases varies.

This phenomenon is encountered, in particular, when a metal electrode is in contact with an electrolyte, which essentially constitutes an interface with a potential difference between the electrode and the electrolyte.

Although the phenomenon may arise no matter what the nature of the two conductive phases present, we shall keep to the case of the ideally-polarizable metal electrode – i.e. without a chemical reaction. The easiest system to study is that of a liquid metal electrode, such as a mercury electrode.

3.8.2. *Gibbs–Lippmann formula and Lippmann's formula*

Consider the ensemble formed of a mercury electrode (the α phase) in contact with an electrolyte (the β phase) containing M_i ions. The interface is supposed to be flat, or at least with a large radius of curvature, meaning that the pressure values on both sides of the interface can be treated as identical. The intervention of the surface and electrical phenomena introduce the $\sigma \mathrm{d}A$ (σ is the electrolyte/electrode interface tension) and $\sum_j \tilde{\mu}_j \,\mathrm{d}n_j$ corresponding amounts of work into the internal energy, whose differential is then written as:

$$\mathrm{d}U = T\mathrm{d}S - P\mathrm{d}V + \sigma\mathrm{d}A + \sum_j \tilde{\mu}_j \,\mathrm{d}n_j \qquad [3.55]$$

We have directly introduced the work of the electrical forces in the form of the electrochemical potentials of the species j. The species to be taken into account are, obviously, the M_i ions of the electrolyte, the solvent and the electrons from the metal. The metal atoms are not taken into account, because the metal only serves as an electron source.

Similarly, we would write the differential of the Helmholtz energy in the form:

$$\mathrm{d}F = -S\mathrm{d}T - P\mathrm{d}V + \sigma\mathrm{d}A + \sum_j \tilde{\mu}_j \,\mathrm{d}n_j \qquad [3.56]$$

The differential of the Gibbs energy would then be:

$$\mathrm{d}G = -S\,\mathrm{d}T + V\,\mathrm{d}P + \sigma\,\mathrm{d}A + \sum_j \tilde{\mu}_j\,\mathrm{d}n_j \qquad [3.57]$$

and we would define a generalized Gibbs energy $\tilde{G}_\sigma$, known as the *electrocapillary Gibbs energy*, whose differential would be:

$$\mathrm{d}\tilde{G}_\sigma = -S\,\mathrm{d}T + V\,\mathrm{d}P - A\,\mathrm{d}\sigma + \sum_j \tilde{\mu}_j\,\mathrm{d}n_j \qquad [3.58]$$

The differential of the surface tension is linked to the excess surface of the components and to the electrochemical potentials by the relation:

$$\mathrm{d}\sigma = -\sum_j \Gamma_j\,\mathrm{d}\tilde{\mu}_j \qquad [3.59]$$

The phase distribution of the j components would then be represented by a curve similar in form to that shown in Figure 3.10. The cross-hatched area on the left of the interphase represents the excess surface of the electrons, Γ_e, and the shaded area to the right of the interphase represents the excess surface Γ_i of a component M_i of the electrolyte.

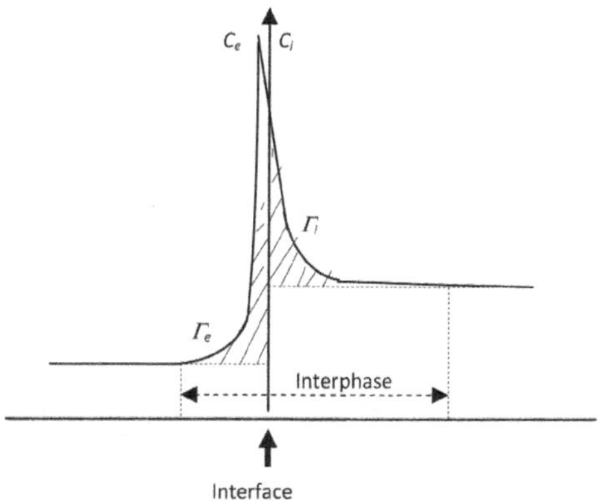

Figure 3.10. *Concentrations of ions and electrons in the electrode/electrolyte interphase*

If z_i represents the number of charges of component M_i, these excess surfaces are linked to one another by the relation of electroneutrality, in the form:

$$\sum_i z_i \Gamma_i = \Gamma_e \quad [3.60]$$

Let us rewrite relation [3.59], separating the terms due to the electrons from those due to the components M_i of the solution. We obtain:

$$\mathrm{d}\sigma = -\sum_i \Gamma_i \,\mathrm{d}\tilde{\mu}_i - \Gamma_e \,\mathrm{d}\tilde{\mu}_e \quad [3.61]$$

The electrochemical potentials are written on the basis of the chemical potentials, in the forms:

$$\begin{cases} \tilde{\mu}_i = \mu_i^{(\beta)} + z_i \mathfrak{F} \varphi^{(\beta)} \\ \tilde{\mu}_e = \mu_e^{(\alpha)} - \mathfrak{F} \varphi^{(\alpha)} \end{cases} \quad [3.62]$$

The chemical potential of the electrons can be deemed to be constant, so by differentiating expressions [3.62], we find:

$$\begin{cases} \mathrm{d}\tilde{\mu}_i = \mathrm{d}\mu_i^{(\beta)} + z_i \mathfrak{F} \,\mathrm{d}\varphi^{(\beta)} \\ \mathrm{d}\tilde{\mu}_e = -\mathfrak{F} \,\mathrm{d}\varphi^{(\alpha)} \end{cases} \quad [3.63]$$

If we take account of the condition of electrical neutrality [3.60], then relation [3.59] becomes:

$$\mathrm{d}\sigma = -\sum_i \Gamma_i \,\mathrm{d}\mu_i - \Gamma_e \mathfrak{F} \left(\varphi^{(\alpha)} - \varphi^{(\beta)} \right) \quad [3.64]$$

The product $\Gamma_e \mathfrak{F}$ represents the surface charge density on the electrode σ_e, and $\left(\varphi^{(\alpha)} - \varphi^{(\beta)} \right)$ is its absolute voltage. In a differential, we can replace the absolute voltage with the relative voltage e, and thus we obtain:

$$\mathrm{d}\sigma = -\sum_i \Gamma_i \,\mathrm{d}\mu_i - \sigma_e \,\mathrm{d}e \quad [3.65]$$

This is the *Gibbs–Lippmann relation*. Employing the hypothesis of constant composition, this formula is simplified to become:

$$\left(\frac{\partial\sigma}{\partial e}\right)_{T,P,\mu_i} = -\sigma_e \qquad [3.66]$$

This latter relation is *Lippmann's formula*.

3.8.3. *Experimentally obtaining the surface tension/electrical potential curve*

The curve showing the variations in surface tension as a function of electrical potential is called the *electrocapillary curve*. To obtain such curves, we use a capillary electrometer (Figure 3.11). A tube, A, containing mercury ends in a capillary tube of diameter r and constitutes an electrode in contact with the electrolyte contained in a tank in which a reference electrode is immersed. A variable voltage is applied between the mercury electrode and a mercury counter-electrode placed at the bottom of the tank. The height h of fluid in the tube of mercury is proportional to the surface tension, and can therefore be used to determine it (see section 4.1.2.2).

3.8.4. *Shape of the electro capillary curves*

If we draw the comparison between the electrode/electrolyte interface with an electrical capacitor, its capacity (which is a positive value) is given by:

$$c = \frac{\mathrm{d}\sigma_e}{\mathrm{d}e} \qquad [3.67]$$

That capacity is constant (independent of the field), and thus by using Lippmann's formula [3.66], we obtain the differential equation:

$$c = -\frac{\mathrm{d}^2\sigma}{\mathrm{d}e^2} \qquad [3.68]$$

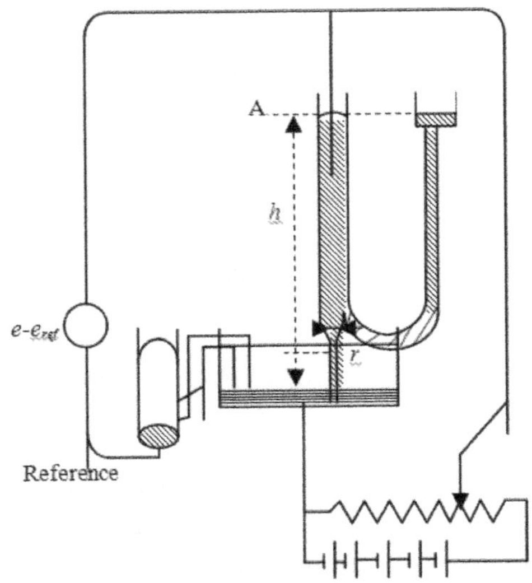

Figure 3.11. *Capillary electrometer*

A first round of integration (where a is a constant) yields:

$$\frac{\mathrm{d}\sigma}{\mathrm{d}e} = -ce + a = -\sigma_s \qquad [3.69]$$

and a second integration (where b is a constant) gives us the equation of the electrocapillary curve:

$$\sigma = -\frac{ce^2}{2} + ae + b \qquad [3.70]$$

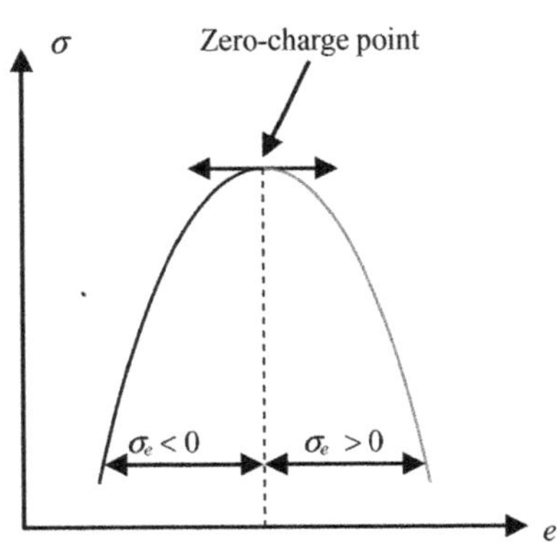

Figure 3.12. *Theoretical electrocapillary curve*

Equation [3.70] shows that the electrocapillary curve is a parabola whose concave surface faces downward ($c > 0$). On the ascending part of the parabola, we can see in view of equation [3.69] that the density of electrical charges is negative, but it is positive on the downward part.

The maximum point of the parabola, where the charge density is 0, is called the *electrocapillary maximum*, and the corresponding tension is the *zero-charge tension*.

Certain authors, in haste, deduced that because the charge density was zero at the maximum point, the same was true of the voltage, in which case it would finally be possible to find the absolute voltage of an electrode. However, experience shows that this is not the case, because the hypothesis of the flat capacitor supposes that electrical charges are only present at the interface, overlooking the presence of dipoles – in particular, the dipoles of the water molecule, which make a notable contribution to the electrode's voltage. Thus, in actual fact, the experimental curves are dissymmetrical (see Figure 3.12), essentially presenting two different parabolic branches. The ascending branch is attributable to the anions. This means that all electrolytes which have the same anion would give approximately the same branch, because the dipole contribution due essentially to water varies little in similar conditions of concentration. The descending curve, on the other hand, is essentially due to the cations, and would be common to practically all electrolytes sharing the same cation.

Figure 3.13 shows the electrocapillary curve of the mercury/iodide system for three iodides: calcium iodide, sodium iodide and potassium iodide. We can see that the ascending parabolic branch is shared by all three curves as, in addition to having the same mercury electrode, the three systems also share the same anion. By contrast, the descending parabolic branches are different, because the cations are different.

NOTE.– It is worth pointing out that the reference electrode used works in the same electrolyte with variable composition as the mercury electrode, so its own absolute voltage is not constant, and the variation measured is not, strictly speaking, that of the voltage of mercury. The electrocapillary curves therefore need to be plotted as a function of the potential E – a potential difference at the terminals (e.g. when using a hydrogen electrode) of the chain:

$$\text{Phases}\ \underbrace{\text{Pt}}_{(1')}\Big|\underbrace{\text{Hg}}_{(1)}\Big|\underbrace{\text{H}_2\text{O}+\text{electrolyte}}_{(2)}\Big|\underbrace{\text{H}_2\,|\text{Pt}}_{(2')} \qquad [3.71]$$

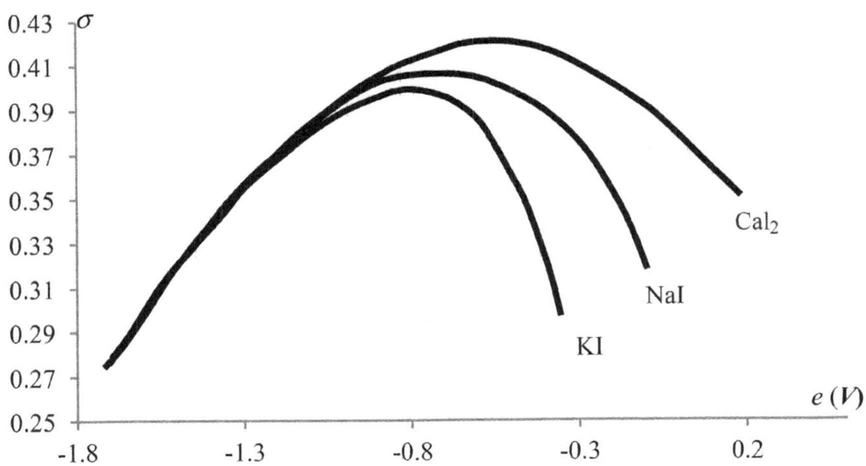

Figure 3.13. *Electrocapillary curves found experimentally*

3.8.5. *Applying electrocapillarity to the experimental determination of the excess surface*

Based on relation [3.65], given to us by Gibbs and Lippmann, we can write the following for the excess surface of component i:

$$\Gamma_i = -\left(\frac{\partial \sigma}{\partial \mu_i}\right)_{T,e,\mu_j} \qquad [3.72]$$

In practice, as we saw for capillary systems (section 2.2.5), the relation existing between the chemical potentials is incapable of delivering the above differentiation. Therefore, all we have access to is the relative adsorptions, in relation to component 1 (e.g. the solvent), which obeys the following relation, derived from expression [2.39]:

$$\Gamma_{i,1} = -\left(\frac{\partial \sigma}{\partial \mu_i}\right)_{T,e,\mu_j} \quad j \neq i \geq 2 \qquad [3.73]$$

Relations [3.72] and [3.73] are only identical for fairly-dilute solutions.

Even using the relative adsorptions or dilute solutions, if component i is an ion, it is not possible to alter its chemical potential without modifying that of the other ions, and thus relations [3.72] and [3.73] need to be modified to show only the chemical potentials of neutral molecules.

To take the example of a simple case, we shall stick with 1:1 electrolytes, such as HCl, which dissociates into Cl^- and H^+ ions, whose relative adsorptions $\Gamma_{H^+,1}$ and $\Gamma_{Cl^-,1}$ we shall now calculate.

To measure the voltage of the mercury electrode used, we pair it with a reference hydrogen electrode, creating the chain in [3.71]. The potential difference at the terminals of that chain can be written as:

$$E = \varphi^{(1')} - \varphi^{(2')} = \left(\varphi^{(1')} - \varphi^{(1)}\right) + \left(\varphi^{(1)} - \varphi^{(2)}\right) + \left(\varphi^{(2)} - \varphi^{(2')}\right) \qquad [3.74]$$

As the potential difference Pt/Hg $\left(\left(\varphi^{(1')} - \varphi^{(1)}\right)\right)$ remains constant regardless of the concentration of the electrolyte, the differential of the potential difference can be written:

$$\mathrm{d}E = \mathrm{d}\left(\varphi^{(1)} - \varphi^{(2)}\right) + \mathrm{d}\left(\varphi^{(2)} - \varphi^{(2')}\right) \qquad [3.75]$$

or

$$\mathrm{d}E = \mathrm{d}e - \mathrm{d}e_{\text{ref}} \qquad [3.76]$$

The voltage e_{ref} of the reference electrode (which, by definition, is unpolarizable) is given by Nernst's relation:

$$\mathrm{d}e_{\text{ref}} = \frac{\mathrm{d}\mu_{\mathrm{H}^+}}{\mathfrak{F}} \qquad [3.77]$$

and thus, in view of relation [3.76], we can write:

$$\mathrm{d}e = \mathrm{d}E + \frac{\mathrm{d}\mu_{\mathrm{H}^+}}{\mathfrak{F}} \qquad [3.78]$$

Hence, in light of the Gibbs–Lippmann relation [3.65], we write:

$$\mathrm{d}\sigma = -\Gamma_{\mathrm{H}^+}\,\mathrm{d}\mu_{\mathrm{H}^+} - \Gamma_{\mathrm{Cl}^-}\,\mathrm{d}\mu_{\mathrm{Cl}^-} - \sigma_s \mathrm{d}E - \sigma_s \frac{\mathrm{d}\mu_{\mathrm{H}^+}}{\mathfrak{F}} \qquad [3.79]$$

The condition of electrical neutrality [3.60] means we must have:

$$\sigma_e = -\left(\Gamma_{\mathrm{H}^+} - \Gamma_{\mathrm{Cl}^-}\right)\mathfrak{F} \qquad [3.80]$$

This relation gives rise to an identical relation between the relative adsorptions:

$$\sigma_e = -\left(\Gamma_{\mathrm{H}^+,1} - \Gamma_{\mathrm{Cl}^-,1}\right)\mathfrak{F} \qquad [3.81]$$

From this, we deduce:

$$\mathrm{d}\sigma = \sigma_e \mathrm{d}E - \Gamma_{\mathrm{Cl}^-,1}\left(\mathrm{d}\mu_{\mathrm{Cl}^-} - \mathrm{d}\mu_{\mathrm{H}^+}\right) \qquad [3.82]$$

By replacing the chemical potentials with their expressions as a function of the activities, we reveal the product of the activities of the chloride ions and hydrogen – i.e. the mean activity of HCl – and thus we have:

$$\mathrm{d}\sigma = \sigma_e \mathrm{d}E - 2RT\Gamma_{\mathrm{Cl}^-,1}\,\mathrm{d}\ln\left|\mathrm{HCl}\right| \qquad [3.83]$$

By plotting the electrocapillary curves $\sigma_s(E)$ for different values of the mean activity of HCl, we can find the relative adsorption of the chloride ions, as follows:

$$\Gamma_{Cl^-,1} = -\frac{1}{2RT}\left(\frac{\partial\sigma}{\partial \ln|HCl|}\right)_{T,P,E} \qquad [3.84]$$

If we also know the surface charge density, we can directly obtain, by finding the difference, the relative adsorption of the protons, in obeisance to the electroneutrality equation, which imposes:

$$\Gamma_{H^+,1} = \Gamma_{Cl^-,1} - \frac{\sigma_s}{\mathfrak{F}} \qquad [3.85]$$

NOTE.– To directly find the relative adsorption of the protons, we must replace the hydrogen electrode with an unpolarizable electrode sensitive to chloride ions – e.g. the calomel electrode.

4

Small-volume Phases

The term *small-dimension phase* denotes phases wherein the dimensions are generally much less than a micrometer. Such phases may include small grains or droplets (small volumes). They are the phases found during the earliest stages of the creation of a liquid or solid condensed phase from a bulk phase in the nucleation process.

Experience shows us that this type of phase exhibits very particular thermodynamic properties, such as a melting point that is not constant during the process of fusion, or the chemical stability of a phase dependent on its dimensions, etc. These special properties are attributed to the relative importance of the surface energies of the interfaces separating those phases from their environment, in comparison to their bulk energy.

In spite of their slight dimensions, these phases tend to be autonomous, which cannot be said of interfaces, whose thermodynamic properties depend – as we saw in the earlier chapters – on those of the adjacent phases.

In this chapter, we shall examine the case of small-volume phases, in the form of spherical liquid drops or small Wulff crystals (see section 3.4).

4.1. Laplace's law for spherical liquid drops

The application of Laplace's law determining the difference in pressure between the inside and the outside of a spherical liquid drop with radius r is written as follows, in view of relations [1.12] and [1.42]:

This and the previous chapter owe a great deal to the kind contribution of Patrice Nortier.

$$p^{(\text{int})} = p^{(\text{ext})} + \frac{2\sigma}{r} \qquad [4.1]$$

This result remains valid for a bubble of gas in a liquid; the difference between the drop and the bubble lies simply in the inversion of the nature of the phases.

4.2. Similarity between the thermodynamics of a Wulff crystal and that of a liquid drop

Let us look again at relation [3.18], which gives the surface energy of a Wulff crystal. Ψ_v and Ψ_s are, respectively, the bulk and surface form factors, and r is the radius of the sphere with the same volume as the crystal:

$$W = 4\pi r^2 \left\{ \Psi_s \left(3\Psi_v\right)^{-2/3} \left(4\pi\right)^{-1/3} \sum_{i=1}^{i_m} \sigma_i \alpha_i \right\} \qquad [4.2]$$

If we define the term in chain brackets as an effective surface tension σ_{eff}, this energy can be written as:

$$W = 4\pi r^2 \sigma_{eff} \qquad [4.3]$$

This relation is absolutely identical to that which gives the surface energy of a spherical liquid drop with radius r given by the application of relation [1.5]. Thus, for a Wulff crystal, we can write the expression of the pressure differential between the inside of the crystal and the fluid (gas or liquid) surrounding it in the form:

$$p^{(\text{int})} = p^{(\text{ext})} + \frac{2\sigma_{eff}}{r} \qquad [4.4]$$

Thus, relations [4.1], pertaining to a spherical liquid drop, and [4.4], pertaining to a Wulff crystal, are absolutely analogous, and therefore we can state that a Wulff crystal is thermodynamically equivalent to a liquid drop.

4.3. Reiss' characteristic function

Consider a small phase composed of a spherical liquid droplet whose volume is $V^{(\text{liq})}$, placed in a closed system with the volume $V^{(\text{gas})}$, containing a gas at the pressure $P^{(\text{ext})}$. The gas is a mixture of the vapor of the liquid and an inert gas which is insoluble in the liquid and not adsorbed to its surface. The temperature of the whole system is kept at the value T.

The function of the generalized potential Gibbs energy of such a system, known as Reiss' function, is written:

$$G_R = U - TS + P^{(ext)}\left(V^{(\text{liq})} + V^{gas}\right) \qquad [4.5]$$

NOTE.– Reiss function is a generalized Gibbs energy function.

As that function is a potential function, any spontaneous transformation at constant temperature T and pressure $P^{(\text{ext})}$ will be characterized by:

$$\left(\mathrm{d}G_R\right)_{T,P^{(\text{ext})}} \leq 0 \qquad [4.6]$$

Equality in the above relation defines the state of equilibrium of the system where Reiss' function exhibits a minimum if that equilibrium is stable.

The internal energy and entropy of that system are linked to the corresponding values characterizing those of the phases present, by the additivity relations:

$$U = U^{(\text{gas})}\left(P^{(\text{ext})}\right) + U^{(\text{liq})}\left(P^{(\text{int})}\right) \qquad [4.7a]$$

$$S = S^{(\text{gas})}\left(P^{(\text{ext})}\right) + S^{(\text{liq})}\left(P^{(\text{int})}\right) \qquad [4.7b]$$

The pressure $P^{(\text{int})}$ is the prevailing pressure inside the droplet.

Let us express Reiss' function on the basis of the internal energies and entropies of each of the phases. We have:

$$\begin{aligned} G_R = {} & U^{(\text{gas})}\left(P^{(\text{ext})}\right) - TS^{(\text{gas})}\left(P^{(\text{ext})}\right) + P^{(\text{ext})}V^{(\text{gas})} \\ & + U^{(\text{liq})}\left(P^{(\text{int})}\right) - TS^{((\text{liq}))}\left(P^{(\text{int})}\right) + P^{(\text{ext})}V^{(\text{liq})} \end{aligned} \qquad [4.8]$$

The sum of the first three terms in this expression represents the Gibbs energy of the gaseous phase at pressure $P^{(\mathrm{ext})}$: $G^{(\mathrm{gas})}\left(P^{(\mathrm{ext})}\right)$. The sum of the last three terms is a different value $J^{(\mathrm{liq})}$ of the Gibbs energy of the drop $G^{(\mathrm{liq})}\left(P^{(\mathrm{int})}\right)$ because of the involvement of the pressure $P^{(\mathrm{ext})}$ in product PV. Relation [4.8] can thus be written in the form:

$$G_R = G^{\mathrm{gas}}\left(P^{(\mathrm{ext})}\right) + J^{(\mathrm{liq})} \qquad [4.9]$$

The value $J^{(\mathrm{liq})}$ here is defined by:

$$\begin{aligned} J^{(\mathrm{liq})} &= U^{(\mathrm{liq})}\left(P^{(\mathrm{int})}\right) - TS^{(\mathrm{liq})}\left(P^{(\mathrm{int})}\right) + P^{(\mathrm{ext})}V^{(\mathrm{liq})} \\ &= F^{(\mathrm{liq})}\left(P^{(\mathrm{int})}\right) + P^{(\mathrm{ext})}V^{(\mathrm{liq})} \end{aligned} \qquad [4.10]$$

Now suppose that the same amount of material as that making up the drop is taken to integrate into the bulk phase by being taken to pressure $P^{(\mathrm{ext})}$. In this operation, $U^{(\mathrm{liq})}\left(P^{(\mathrm{int})}\right)$ tends toward $U^{(\mathrm{liq})}\left(P^{(\mathrm{ext})}\right)$ and $S^{(\mathrm{liq})}\left(P^{(\mathrm{int})}\right)$ tends toward $S^{(\mathrm{liq})}\left(P^{(\mathrm{ext})}\right)$, so $J^{(\mathrm{liq})}$ tends toward $G^{(\mathrm{liq})}\left(P^{(\mathrm{ext})}\right)$ which, if we overlook the compressibility of the liquid, is such that:

$$\begin{aligned} &G^{(\mathrm{liq})}\left(P^{(\mathrm{ext})}\right) = U^{(\mathrm{liq})}\left(P^{(\mathrm{ext})}\right) - TS^{(\mathrm{liq})}\left(P^{(\mathrm{ext})}\right) + P^{(\mathrm{ext})}V^{(\mathrm{liq})} \\ &= F^{(\mathrm{liq})}\left(P^{(\mathrm{ext})}\right) + P^{(\mathrm{ext})}V^{(\mathrm{liq})} \end{aligned} \qquad [4.11]$$

Using relations [4.10] and [4.11], we can calculate the difference:

$$J^{(\mathrm{liq})} - G^{(\mathrm{liq})}\left(P^{(\mathrm{ext})}\right) = F^{(\mathrm{liq})}\left(P^{(\mathrm{int})}\right) - F^{(\mathrm{liq})}\left(P^{(\mathrm{ext})}\right) \qquad [4.12]$$

By finding $J^{(\mathrm{liq})}$ from relation [4.12] and feeding it back into expression [4.9], we obtain:

$$G_R = G^{(\mathrm{gas})}\left(P^{(\mathrm{ext})}\right) + G^{(\mathrm{liq})}\left(P^{(\mathrm{ext})}\right) + F^{(\mathrm{liq})}\left(P^{(\mathrm{int})}\right) - F^{(\mathrm{liq})}\left(P^{(\mathrm{ext})}\right) \qquad [4.13]$$

The difference $F^{(\text{liq})}\left(P^{(\text{int})}\right)-F^{(\text{liq})}\left(P^{(\text{ext})}\right)$ represents the reversible isothermal work needed for the transfer of a certain quantity of liquid, corresponding to that of the future drop, at pressure $P^{(\text{ext})}$, to form the droplet with radius r at pressure $P^{(\text{int})}$. This transfer can be enacted in an experiment, represented by Figure 4.1, where the liquid initially forming a bulk phase is placed in a syringe. The pressure exerted on piston p_2 of the syringe causes the formation of the drop with radius r and volume $V^{(\text{liq})}$ at the end of the needle.

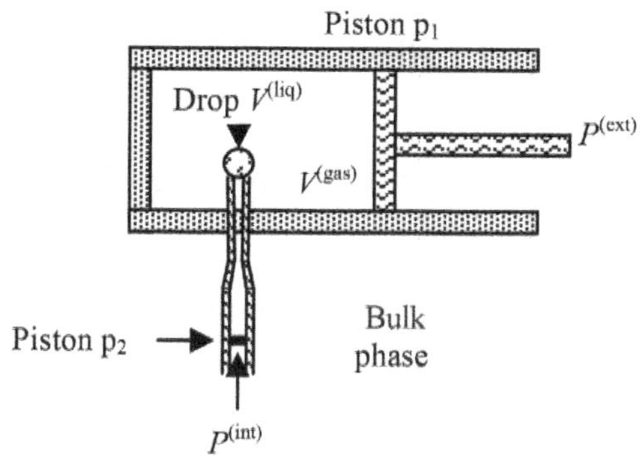

Figure 4.1. *Formation of a drop at a constant imposed external pressure*

When the drop is at mechanical equilibrium, Laplace's law [4.1] must be satisfied. The total work necessary to form the drop is:

$$W_{\text{tot}}=\int_0^r\left(P^{(\text{ext})}+\frac{2\sigma}{r}\right)4\pi r^2\,\mathrm{d}r \qquad [4.14]$$

The work found on the left-hand side of this equation, $\int_0^r 4\pi r^2 P^{(\text{ext})}\,\mathrm{d}r$, is, in fact, restored to the outside by the motion of the piston p_1 in the cylinder in order to keep the pressure $P^{(\text{ext})}$ constant. Thus, the only portion of the work which contributes to the increase in the liquid's Helmholtz energy is:

$$W=F^{(\text{liq})}\left(P^{(\text{int})}\right)-F^{(\text{liq})}\left(P^{(\text{ext})}\right)=\sigma\int_0^r 8\pi r\,\mathrm{d}r \qquad [4.15]$$

In light of relation [4.13], we then find:

$$G_R = G^{(\text{gas})}\left(P^{(\text{ext})}\right) - G^{(\text{liq})}\left(P^{(\text{int})}\right) + 4\pi r^2 \sigma \qquad [4.16]$$

Hence, we know the value of Reiss' function for a spherical liquid drop. We can see that the variables in that function are temperature T, internal pressure $P^{(\text{int})}$, external pressure $P^{(\text{ext})}$ and the radius of the drop r.

In view of the equivalence shown in section 4.1.2, expression [4.16], which was established for a spherical liquid drop, is also applicable to a Wulff crystal with equivalent radius (see section 3.4).

4.4. Gibbs energy of a spherical pure liquid or solid with small volume

Once again, let us consider a spherical liquid drop with the radius r.

As the gaseous phase is a mixture of the vapor of the liquid and an inert gas, the outside pressure is the sum of the partial pressure of the inert gas, P_g, and the pressure of the liquid vapor P_v, so:

$$P^{(\text{ext})} = P_g + P_v \qquad [4.17]$$

If n_v and n_g are, respectively, the quantities of vapor and of inert gas, the Gibbs energy of the gaseous phase will be:

$$G^{(\text{gas})}\left(P^{(\text{ext})}\right) = n_g g_g^0\left(P_g\right) + n_v g_v^0\left(P_v\right) \qquad [4.18]$$

If the drop of liquid contains n_l moles, its Gibbs energy will be:

$$G^{(\text{liq})}\left(P^{(\text{ext})}\right) = n_l g_l^0\left(P^{(\text{ext})}\right) \qquad [4.19]$$

The material balance means that we can write the following relation between the variations in the quantity of liquid and vapor:

$$\mathrm{d}n_v = -\mathrm{d}n_l \qquad [4.20]$$

By differentiating relation [4.16] in relation to the quantity n_l, we can write:

$$(\mathrm{d}G_R)_{T,P^{(\mathrm{ext})}} = \frac{\partial G_R}{\partial n_l} = \left[g_l^0\left(P^{(\mathrm{ext})}\right) - g_v^0\left(P_v\right) \right] \mathrm{d}n_l + 8\pi r\sigma \left(\frac{\partial r}{\partial n_l} \right) \mathrm{d}n_l = 0 \quad [4.21]$$

Thus, we obtain the relation between the molar Gibbs energies of the liquid and of the vapor:

$$g_v^0\left(P_v\right) = g_l^0\left(P^{(\mathrm{ext})}\right) + \frac{2\sigma}{r} v^{0(\mathrm{liq})} \quad [4.22]$$

This relation, in view of expression [4.1], can be written:

$$g_v^0\left(P_v\right) = g_l^0\left(P^{(\mathrm{ext})}\right) + \left(P^{(\mathrm{int})} - P^{(\mathrm{ext})}\right) v^{0(\mathrm{liq})} \quad [4.23]$$

We know that in the case of a massive liquid phase, the Gibbs energy of the vapor is equal to that of the liquid – i.e. a liquid with infinite radius $g_l^0\left(r = \infty\right)$. Hence, for a pure substance A, the Gibbs energy of the liquid phase will be:

$$g_A^0\left(r\right) = g_A^0\left(r = \infty\right) + \frac{2\sigma}{r} v_A^{0(\mathrm{liq})} \quad [4.24]$$

This is the *Gibbs–Thomson relation for pure substances of small volume*.

Of course, this relation can also be applied to a Wulff crystal at equilibrium with its vapor or its pure liquid, with equivalent radius *r*.

4.5. Chemical potential of a component of a solution

We shall now consider the case where the liquid and vapor are polycomponent phases. By differentiating expression [4.16] in relation to the amount of one of the components, *i*, and making that expression equal to 0 (equilibrium), we find:

$$\left(\frac{\partial G_R}{\partial n_i} \right)_{n_j} = \left[\mu_i\left(P^{(\mathrm{ext})}\right) - \mu_i\left(P_v\right) \right] \mathrm{d}n_i + 8\pi r\sigma \left(\frac{\partial r}{\partial n_i} \right) \mathrm{d}n_i = 0 \quad [4.25]$$

This relation gives us:

$$\mu_i^0(r) = \mu_i^0(r = \infty) + \frac{2\sigma}{r}\bar{V}_i \qquad [4.26]$$

Note that relation [4.26], at equilibrium, expresses the equality of the generalized chemical potential of component *i* in the liquid phase and the chemical potential of component *i* in the gaseous phase.

Relation [4.26] is the *Gibbs–Thomson relation for species in solution in small-volume phases*.

NOTE.– Reiss pointed out that relation [4.26] was incorrect, because in the differentiation leading to relation [4.21], for small objects, it was not possible to vary the amount of one of the components and still preserve the composition of the phase. In the case of a binary solution formed of components 1 and 2, an additional term needs to be brought into play, and thus for component 1, the exact relation would be:

$$\mu_1^0(r) = \mu_1^0(r = \infty) + \frac{2\sigma}{r}\bar{V}_1 + \frac{3V_m^{(\text{liq})}(1 - x_1)}{r}\frac{\partial \sigma}{\partial x_1} \qquad [4.27]$$

where $V_m^{(\text{liq})}$ is the molar volume of the solution.

The additional term only becomes significant if component 1 is dilute and is tensio-active ($\frac{\partial \sigma}{\partial x_1} < 0$), so it is the opposite of the increase in chemical potential due to the Gibbs–Thomson effect.

4.6. Phase change in pure substances

We shall now examine the impact of small radii of curvature on the thermodynamic properties of the phase transformations of pure substances, using the examples of the vaporization of a drop of liquid and the melting of a solid made up of small grains.

4.6.1. *The saturating vapor pressure of pure liquid*

Consider a liquid droplet of a pure substance, with radius r at equilibrium with its vapor at a given temperature T. The total pressure of the vapor phase is kept constant by an inert gas. If equilibrium is achieved, the generalized chemical potentials of the two phases are equal. With the notations used for phase changes, relation [4.22] gives us:

$$g^{0(\text{vap})} = g^{0(\text{liq})} + \frac{2\sigma}{r} v^{0(\text{liq})} \quad [4.28]$$

The exchange of an infinitesimal amount between the liquid and the vapor leads to a variation of the chemical potential of the vapor which, supposing the surface tension to be independent of the radius of the drop, is written by differentiation:

$$\left(\mathrm{d}\, g^{0(\text{vap})}\right)_{T,P^{ext}} = 2\sigma v^{0(liq)}\, \mathrm{d}\left(\frac{1}{r}\right) \quad [4.29]$$

In addition, if the gaseous mixture is supposed to be perfect, the infinitesimal variation of the chemical potential of the vapor can thus be written:

$$\left(\mathrm{d}\, g^{0(\text{vap})}\right)_{T,P^{ext}} = v^{0(vap)}\, \mathrm{d}\, P^{vap} = \frac{\mathrm{R}T\, \mathrm{d}\, P^{vap}}{P^{vap}} \quad [4.30]$$

By making expressions [4.29] and [4.30] equal and integrating between the radii r and infinity, for saturating vapor pressures lying between $P^{(\text{vap})}(r)$ and $P^{0(\text{vap})}$, we find:

$$P^{(\text{vap})}(r) = P^{0(\text{vap})} \exp\left(\frac{2\sigma v^{0(\text{liq})}}{\mathrm{R}Tr}\right) \quad [4.31]$$

$P^{0(\text{vap})}$ is the saturating vapor pressure when the surface of the liquid is flat – i.e. the saturating vapor pressure of a typical bulk phase. Relation [4.31] is known as *Kelvin's formula*.

If the radius is not too small, the term beneath the exponential is slight, and if we keep only the first two terms in the limited expansion of the exponential, relation [4.31] can be simplified to give:

$$\frac{P^{(\text{vap})}(r)-P^{0(\text{vap})}}{P^{0(\text{vap})}} \approx \frac{2\sigma v^{0(\text{liq})}}{\text{R}Tr} \qquad [4.32]$$

Thus, the saturating vapor pressure of the drop with radius r is greater than the liquid's normal saturating vapor pressure.

In order for the influence of the small dimension to become apparent, we need to use drops with a very small radius. The second column in Table 4.1 gives a few values of the ratio of the saturating pressure above the drop to the saturating vapor pressure above a flat liquid for different drop radius values. These values are calculated, using relation [4.31], for water with a surface tension of $\sigma = 0.073\text{J/m}^2$ at 15°C.

r (m)	Drop $P^{(\text{vap})}(r) / P^{0(\text{vap})}$	Bubble $P^{(\text{vap})}(r) / P^{0(\text{vap})}$
∞	1	1
10^{-6}	1.001	0.9990
10^{-7}	1.011	0.9891
10^{-8}	1.115	0.897

Table 4.1. *Influence of the radii of drops and bubbles on the saturating vapor pressure of water at 15°C*

Thus, the liquid–vapor equilibrium, which is monovariant for a liquid with a flat surface, becomes divariant for drops of small dimensions. The saturating vapor pressure becomes a function of the temperature (particularly because of pressure $P^{0(\text{vap})}$) and of the radius of the drops.

We can show that, at constant saturating vapor pressure, the smaller the drop is, the lower the equilibrium temperature will be.

NOTE.– It is easy to show that a set of liquid droplets constitutes an unstable system. Indeed, any fluctuation in the size of one of the droplets will cause a variation in the saturating vapor pressure in its vicinity, giving rise to a vapor gradient between the vicinity of that drop and the vicinity of the other drops.

This gradient leads to diffusion, which breaks the equilibrium in the vicinity of the other drops. The equilibrium can only be re-established by the variation of the dimensions of the other drops. Thus, little by little, the process will lead to the disappearance of all but one of the drops, with that one gaining in size at the expense of the others.

The same reasoning as that which yielded relation [4.31] can be employed if we consider not a drop of liquid in a gas, but a bubble of gas in liquid. The saturating vapor pressure then becomes as follows, with the gas being on the convex side of the curvature:

$$P^{(\mathrm{vap})}(r) = P^{0(\mathrm{vap})} \exp\left(-\frac{2\sigma v^{0(\mathrm{liq})}}{\mathrm{R}Tr}\right) \qquad [4.33]$$

Thus, the saturating vapor pressure in a bubble within a liquid is lower than the vapor pressure of the flat liquid. For spheres, the term *Kelvin radius*, r_K, denotes the radius at equilibrium with a certain vapor pressure $P^{(\mathrm{vap})}(r_K)$.

The third column in Table 4.1 shows a few values of the ratio between the pressures in the case of an air bubble of different radii in water.

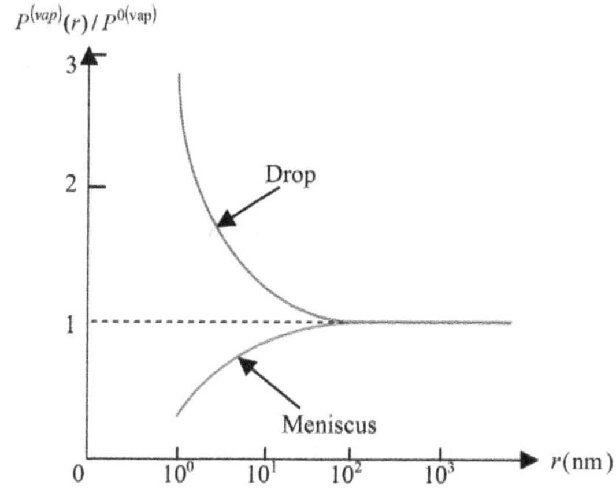

Figure 4.2. *Condensation in the vicinity of a spherical surface: concave (for a drop) or convex (for a meniscus)*

Figure 4.2 shows the results of relations [4.31] and [4.33] on condensation in the vicinity of a spherical surface that is convex (drop) or concave (meniscus or bubble).

4.6.2. *Melting of a small grain*

In the same way as we saw in the case of the liquid–vapor equilibrium for a drop of liquid, the variance of the solid–liquid equilibrium in a small solid grain will increase by one unit, the melting point will depend on the radius of the grains and therefore will not be constant throughout the process of melting of the grain.

In modeling fusion (melting), we shall accept the hypothesis that a solid spherical grain gives rise to a spherical drop of liquid, and that the respective radii of those two spheres are linked, because of the conservation of matter, to the molar volumes of the two phases, thus:

$$\left(\frac{r^{(\text{liq})}}{r^{(\text{sol})}}\right)^3 = \left(\frac{v^{0(\text{liq})}}{v^{0(\text{sol})}}\right) \quad [4.34]$$

We suppose that the substrate on which the particle rests is not wetted by the liquid (link angle of π).

The Gibbs energies (i.e. the generalized Gibbs energies which involve the surface energy or Reiss functions) of the two phases at the melting point T_f, with the liquid being at the internal pressure $P^{(\text{liq})}$ and the solid at internal pressure $P^{(\text{sol})}$ are equal when the crystal and liquid are at equilibrium, so:

$$g_\sigma^{0(\text{liq})}\left(P^{(\text{liq})}, T_f\right) = g_\sigma^{0(\text{sol})}\left(P^{(\text{sol})}, T_f\right) \quad [4.35]$$

If we apply the Gibbs–Thomson relation [4.24] for the solid and liquid and feed the expressions thus obtained back into relation [4.35], we obtain:

$$g^{0(\text{liq})}\left(P^{(\text{ext})}, T_f\right) + \frac{2v^{0(\text{liq})}\sigma^{(\text{liq})}}{r^{(\text{liq})}} = g^{0(\text{sol})}\left(P^{(\text{ext})}, T_f\right) + \frac{2v^{0(\text{sol})}\sigma^{(\text{sol})}}{r^{(\text{sol})}} \quad [4.36]$$

$v^{0(\text{liq})}, v^{0(\text{sol})}$, $\sigma^{(\text{liq})}$ and $\sigma^{(\text{sol})}$ respectively denote the molar volumes and the surface tensions of the liquid phase and the solid phase. The pressure $P^{(\text{ext})}$ is the reference external pressure, usually chosen as 1 bar.

The Gibbs energy of melting of a grain of infinite dimensions $\Delta_f G\left(P^{(\text{ext})}, T_f\right)$ is given by the difference between the chemical potentials:

$$\begin{aligned} & g^{0(\text{liq})}\left(P^{(\text{ext})}, T_f\right) - g^{0(\text{sol})}\left(P^{(\text{ext})}, T_f\right) = \Delta_f G\left(P^{(\text{ext})}, T_f\right) \\ & = \Delta_f H\left(P^{(\text{ext})}, T_f\right) - T_f \Delta_f S\left(P^{(\text{ext})}, T_f\right) \end{aligned} \qquad [4.37]$$

Taking account of relation [4.34], relation [4.36] gives us:

$$\Delta_f H - T_f \Delta_f S + \frac{2v^{0(\text{sol})}}{r^{(\text{sol})}} \left[\sigma^{(\text{liq})} \left(\frac{v^{0(\text{liq})}}{v^{0(\text{sol})}} \right)^{2/3} - \sigma^{(\text{sol})} \right] = 0 \qquad [4.38]$$

In view of the slight differences in temperature, we can overlook the variations, with temperature, of the molar volumes, the surface tensions, the enthalpy of fusion and the entropy of fusion. The new melting point of the grains with radius r can then be calculated as a function of the melting point for grains with an infinite radius T_f^∞, and we obtain:

$$T_f^\infty - T_f = \frac{2v^{0(\text{sol})}}{\Delta_f S} \left[\sigma^{(\text{sol})} - \sigma^{(\text{liq})} \left(\frac{v^{0(\text{liq})}}{v^{0(\text{sol})}} \right)^{2/3} \right] \frac{1}{r^{(\text{sol})}} \qquad [4.39]$$

Thus, the difference between the melting points of a massive solid and of a small grain is approximately inversely proportional to the radius of the grains of the solid. Figure 4.3 shows the variation in the melting points of spherical grains of gold as a function of their radius ($\Delta_f S$ = 9.39J.mol^{-1}; $\sigma^{(\text{liq})}$ = 1.138J.m^{-2}; $\sigma^{(\text{sol})}$ = 1.400J.m^{-2}; $v^{0(\text{liq})}$ = 1.138×10^{-5}m^3.mol^{-1}; $v^{0(\text{sol})}$ = 1.074×10^{-5}m^3.mol^{-1}).

Hence, the effect of dimension is noticeable only in the case of grains whose radius is around a few nanometers.

This effect of dimension can be generalized to all phase-changes in pure substances, and we find the same result: the variance of the transformation is increased by one unit. For instance, zirconia has two polymorphic varieties: the monoclinical variety, which is stable at low temperatures; and the quadratic variety, which is stable at higher temperatures. We can see that for smaller grains, the quadratic phase is stabilized at a lower temperature, or, put differently, the temperature of the polymorphic transformation is a function of the grain dimension, with a curve which is of the same form as that shown in Figure 4.3. Furthermore, relation [4.39] is valid for the temperature of that transformation if we replace the values pertaining to the solid and the liquid with the same values pertaining to the monoclinical and quadratic polymorphic varieties.

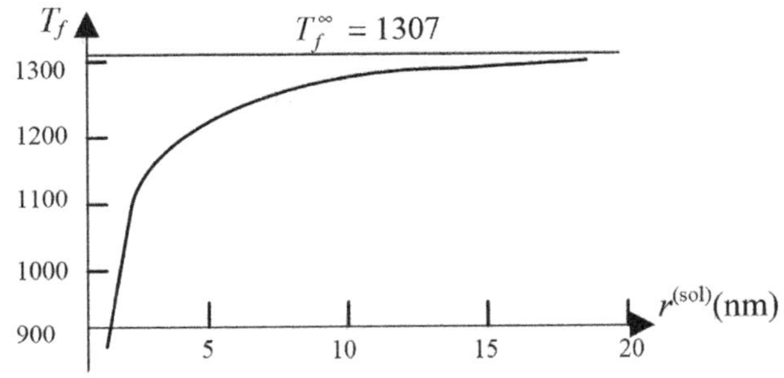

Figure 4.3. *Influence of radius on the melting point of spherical particles of gold*

4.7. Alteration of the solubility of a solid due to the small dimension of its grains

We shall now show that, in the same way as for the equilibria of transformation of pure substances, the variance of the equilibriums of phase transformations in solution is also increased by one unit by the involvement of the dimensions of the phases. To illustrate this, we shall consider the solubility of a solid with small dimensions – i.e. the equilibrium between a pure solid of small volume and a solution. Let us consider a liquid solution with infinite radius of curvature, containing multiple components, including component i. This solution is at equilibrium with a small crystal of the pure substance i with equivalent radius r. We shall let $\sigma^{(\mathrm{ls})}$ denote the interfacial energy between the solution and the crystal whose molar volume is $v^{0(\mathrm{s})}$. The molar fraction of component i in the solution is x_i, and its activity $a_i = \gamma_i x_i$. Its chemical potential is $\mu_i\,(x_i,T)$. The molar Gibbs energy of i in the solid is: $g_i^{0(\mathrm{s})}(r,T)$. At equilibrium, the following relation is satisfied:

$$g_i^{0(\mathrm{s})}(r,T) = \mu_i\,(x_i,T) \qquad [4.40]$$

A shift of the equilibrium at constant temperature would be expressed by the relation:

$$\mathrm{d}\,g_i^{0(\mathrm{s})}(r,T) = \mathrm{d}\mu_i\,(x_i,T) \qquad [4.41]$$

The differential, at constant temperature, of the chemical potential of component i in the solution is given by:

$$\mathrm{d}\mu_i\,(x_i,T) = \mathrm{R}Td\ln a_i \qquad [4.42]$$

For the small crystal, the application of Laplace's relation, where $P^{(\mathrm{int})}$ is the crystal's internal pressure, gives us:

$$\left(\mathrm{d}\,g_i^{0(\mathrm{s})}\right)_T = v^{0(\mathrm{s})}\,\mathrm{d}\,P^{(\mathrm{int})} = 2v^{0(\mathrm{s})}\,\mathrm{d}\left(\frac{\sigma^{(\mathrm{ls})}}{r}\right) \qquad [4.43]$$

If $x_i^{(\infty)}$ and $\gamma_i^{(\infty)}$, respectively, represent the molar fraction and the activity coefficient of component i in the liquid solution at equilibrium with a solid i of infinite radius, then by substituting relations [4.42] and [4.43] back into expression [4.41] and integrating between the radii r and ∞, we find:

$$\gamma_i x_i(r) = \gamma_i^{(\infty)} x_i^{(\infty)} \exp\left(\frac{2v^{0(\mathrm{s})}\sigma^{(\mathrm{ls})}}{\mathrm{R}T}\frac{1}{r}\right) \qquad [4.44]$$

If, in addition, the solution is sufficiently dilute, we can add the hypothesis that $\gamma_i^{(\infty)} = \gamma_i$. Expression [4.44] is then simplified:

$$x_i(r) = x_i^{(\infty)} \exp\left(\frac{2v^{0(s)}\sigma^{(ls)}}{RT}\frac{1}{r}\right) \qquad [4.45]$$

Thus, the solubility of the solid is inversely proportional to the radius of the particles.

The calculation shows that in the case of a crystal with radius 100 nm, the solubility is increased by around 4% (with $\sigma^{(ls)} = 0.5\ \mathrm{J.m^{-2}}$, at 300 K for a solid with the molar volume $10^{-5}\ \mathrm{m^3.mol^{-1}}$), in comparison to the solubility of a large crystal.

NOTE.– As in the case of the liquid–vapor equilibrium, if a solution contains a collection of small solid grains, because the smallest particles are the most soluble, they create a gradient of concentration in the liquid between the vicinity of the smaller and larger particles. This gradient causes diffusion, which causes the largest particles to grow; the system is not stable and evolves toward a single particle with a large radius, which is known as *Ostwald ripening*; the small particles disappear and amalgamate with the larger ones.

4.8. Equilibrium constant for a reaction involving small grains

The influence of the radius of curvature of the phase on the Gibbs energy or the chemical potential of a component in that phase leads to the alteration of the values of the equilibrium constants every time a chemical reaction takes place with phases of small dimensions. We shall discuss the formation of nickel carbonyl in light of Mittasch's experiments, interpreted by Defay and Prigogine (1951) as an example.

The reaction of solid nickel with gaseous carbon monoxide to give nickel carbonyl, also in the gaseous state, at temperature T (70°C), is thus:

$$\langle \mathrm{Ni} \rangle + 4\{\mathrm{CO}\} = \{\mathrm{Ni(CO)_4}\} \qquad [4R.1]$$

The components Ni, CO and $Ni(CO)_4$ will be spoken of as components 1, 2 and 3, respectively.

If σ is the surface energy of the nickel in the presence of the gaseous phase, the expression of the Reiss potential at temperature T and pressure $P^{(\text{ext})}$ is:

$$G_R = n_1\mu_1\left(P^{(\text{ext})}\right) + n_2\mu_2\left(P^{(\text{ext})}\right) + n_3\mu_3\left(P^{(\text{ext})}\right) + 4\pi r^2\sigma \tag{4.46a}$$

The differential of this function at constant temperature and external pressure is, in view of the Gibbs–Duhem relation for the gaseous phase:

$$\left(\mathrm{d}G_R\right)_{T,P^{(\text{ext})}} = \mu_1\,\mathrm{d}n_1 + \mu_2\,\mathrm{d}n_2 + \mu_3\,\mathrm{d}n_3 + 8\pi r\sigma\frac{\partial r}{\partial n_1}\mathrm{d}n_1 \tag{4.46b}$$

At equilibrium, this differential is zero:

$$\left(\mathrm{d}G_R\right)_{T,P^{(\text{ext})}} = 0 \tag{4.47}$$

The stoichiometry of the reactions imposes the relations $\mathrm{d}n_2 = 4\mathrm{d}n_1$ and $\mathrm{d}n_3 = -\mathrm{d}n_1$ between the variations in the amounts of matter, and the spherical form means we can write:

$$\frac{\partial r}{\partial n_1} = \frac{v_1^0}{4\pi r^2} \tag{4.48}$$

The equilibrium condition then becomes:

$$\mu_1(\infty) + 4\mu_2 - \mu_3 = \frac{2v_1^0\sigma}{r} \tag{4.49}$$

At constant pressure and temperature, for nickel we have $\mathrm{d}\mu_1(\infty) = 0$ because the phase is pure. Then, by differentiating relation [4.49], we obtain:

$$4\,\mathrm{d}\mu_2 - \mathrm{d}\mu_3 = -2v_1^0\sigma\,\mathrm{d}\left(\frac{1}{\mathrm{r}}\right) \tag{4.50}$$

If the gaseous mixture constitutes a perfect solution, for $i = 2$ and 3 we have:

$$\mathrm{d}\mu_i = \mathrm{R}T\,\mathrm{d}\ln x_i \qquad [4.51]$$

By substituting back into relation [4.50] with the condition $x_2 = 1 - x_3$, we find:

$$\frac{1+3x_3}{1-x_3}\frac{\mathrm{d}x_3}{x_3} = \frac{2v_1^0\sigma}{\mathrm{R}T}\mathrm{d}\left(\frac{1}{r}\right) \qquad [4.52]$$

Let $x_3(r)$ denote the molar fraction of nickel carbonyl in the gas at equilibrium with a spherical metal sample of radius r, and $x_3(\infty)$ the same fraction at equilibrium with a flat sample of nickel. By integrating the expression [4.52] between the two radii $r = r$ and $r = \infty$ (planar liquid), we obtain:

$$\ln\frac{x_3(r)}{x_3(\infty)} - 4\ln\frac{1-x_3(r)}{1-x_3(\infty)} = \frac{2v_1^0\sigma}{\mathrm{R}T}\frac{1}{r} \qquad [4.53]$$

We express the equilibrium constants for the two radii, r and ∞, by:

$$K(r) = \frac{x_3(r)}{\left[1-x_3(r)\right]^4}\left(P^{(\text{ext})}\right)^3 \qquad [4.54a]$$

and

$$K(\infty) = \frac{x_3(\infty)}{\left[1-x_3(\infty)\right]^4}\left(P^{(\text{ext})}\right)^3 \qquad [4.54b]$$

We find a relation between the equilibrium constant obtained for a metal sample with radius r and the equilibrium constant for a flat metal sample:

$$K(r) = K^{\infty}\exp\frac{2v_1^0\sigma}{\mathrm{R}T}\frac{1}{r} \qquad [4.55]$$

Thus, in our example, at constant temperature and pressure, the equilibrium constant and the progress of the reaction toward equilibrium increase if the radius of the solid reagent decreases. Note the similarity between relations [4.31] and [4.55] – a similarity which fits with the fact that the vapor pressure at equilibrium is an equilibrium constant. In the same conditions as for equation [4.31], equation [4.55] gives us a relation similar to expression [4.32]: thus, the equilibrium constant is essentially inversely proportional to the radius of the grains.

4.9. Nucleation of a condensed phase

The nucleation of a condensed phase is the process whereby, from a different phase (fluid or solid), tiny fragments of that new phase appear. This process is a phenomenon which can be modeled by using the thermodynamic properties of the system formed of the two phases, the newer of which has extremely small dimensions.

It is recognized that the driving force behind the creation of a solid is oversaturation, and that the aggregates which are the inevitable intermediary products correspond to states of energy higher than those of the infinite solid. The oversaturation and the interfacial energy will be the two main terms involved in the creation of nuclei.

We shall adopt the classic terminology: the aggregates make up the reaction path. Nucleation from a system that was initially free of the phase formed is known as primary nucleation. If it occurs within an initial phase, it is said to be homogeneous, whether in a fluid or solid medium. Nucleation which takes place upon contact with a solid phase is said to be heterogeneous.

A nucleus of a condensed phase contains a number N of molecules, and is produced by the condensation of those N molecules belonging to a precursor.

4.9.1. *Hypotheses underlying the nucleation model*

When attempting to establish the Reiss function of formation of nuclei from the precursor or the Gibbs energy of condensation, we come up against a difficulty pertaining to the definition of the reference state:

– the newly-condensed phase forms a pure phase, which therefore constitutes the reference state. The chemical potential of the single component of that phase is constant and equal to its molar Gibbs energy of formation;

– on the other hand, species A (the precursor) is present in a phase with multiple components (molecules or ions of a fluid or point defects of a solid), its reference state depends on the choice of convention (I, II or III), and its chemical potential depends on its abundance, expressed as a molar fraction, for example;

– the aggregates are in an intermediary situation: it is tempting to consider the smallest entities as being oligomers, small molecules in solution whose chemical potential varies with concentration. This choice is in line with the origin of the term of the molar fraction (or concentration) in the chemical potential: it stems essentially from the entropy of mixing. Whilst it is logical to consider that the juxtaposition of an *infinite* condensed phase which decants and a solution floating upon it does not create disorder, we are forced to make the opposite observation in the case of small particles kept in suspension by Brownian motion. In addition, the application of the law of mass action to aggregates considered as pure phases whose chemical potential is independent of the composition would lead us to conclude that for each oversaturation, there is a corresponding nucleus size at equilibrium, which is belied by real-world experience.

Thus, we need to treat the aggregates as species dissolved in the initial phase. Hence, we introduce a discontinuity between the aggregates of increasing size and the new condensed phase (in the form of a nucleus) although, for the moment, we are unable to define a critical size at which that discontinuity would occur.

We now need to evaluate the partial molar Gibbs energy of the aggregates in the reference state: let us base our reasoning on convention I, so the reference state of any component is the pure state. Consider the hypothetical system in which each aggregate is surrounded by a layer of the initial phase, of infinitesimal thickness, but nonetheless sufficient for the surface energy term to be established. The molar fraction of the aggregates, then, is practically equal to 1, and their chemical potential is that of their reference state. This situation is almost identical to that of the condensed phase making up the nucleus, with the exception of the surface Gibbs energy term:

$$\mu_{R2,N}^{\circ(I)} = \frac{N}{N_a} \cdot g_{R2_c}^{0} + \Delta_{R2,N}(\sigma A) \qquad [4.56]$$

Convention (I) is not well suited to dilute solutions, for which we tend to use references (II) or (III). It is not possible to keep the activity coefficient at the value of 1 unless the solution is perfect.

Let $\Delta_{R2,N}G$ be Reiss' standard Gibbs energy of condensation associated with the reaction of formation of a nucleus of N construction units, from the initial phase – i.e. the reaction:

$$N\text{A} = \text{A}_N \qquad [4R.2]$$

For the variation of the Reiss potential, in view of relation [4.16], we have:

$$\Delta_{R2,N}G_R = N.(g_A^0 - \mu_A^0) + RT\ln|A| + \sigma_L.\Delta_{R2,N}(A) \qquad [4.57]$$

The term $\Delta_{R2,N}(A)$ denotes the variation in surface energy between the nucleus and the initial state of the system, all at constant external pressure $P^{(ext)}$, temperature T and N. σ_L is the interfacial energy between the initial phase (liquid or gaseous) and the final condensed phase, be it solid or liquid.

In view of the variation of the molar Gibbs energy associated with the transformation [4R.2], which is written:

$$\Delta_{R2}g = \left(g_A^0 - \mu_A^0\right) + RT\ln|A| \qquad [4.58]$$

If N_a is Avogadro's number, the variation of the Reiss potential function of the nucleation is:

$$\Delta_{R2,N}G_R = \frac{N}{N_a}\Delta_{R2}g + \sigma_L\Delta_{R2,N}(A) \qquad [4.59]$$

It is only possible to process this relation if we hypothesize that the interfacial energy is not dependent upon the size of the aggregate. There is no theoretical justification for this approximation when the aggregates are very small (less than a thousand construction units); on the other hand, it is logical to overlook the effects of edges or quantum effects on larger objects.

4.9.2. *Homogeneous nucleation in a fluid phase: Volmer's approach (1905)*

In the case of nucleation of a condensed phase (solid or liquid) from a fluid phase (liquid or gaseous), we can state that:

– if the condensed phase is liquid, the initial phase is gaseous and, given what we already know (section 4.2), the nucleus is spherical in shape (equilibrium of shape of a liquid drop);

– if the nucleus formed is solid, then the primary phase is liquid or gaseous, but it is always possible to treat it from the point of view of a spherical equivalent nucleus, by using an effective interfacial energy for the solid.

In view of the above hypotheses, the volume of the aggregate can be obtained either as the volume of a sphere with radius r, or on the basis of the molar volume v_L^0 (because the aggregate is a *portion* of condensed phase, its compressibility is taken to be zero), and the two values must be equal, so:

$$\frac{4\pi}{3} r^3 = N.\frac{v_L^0}{\mathrm{N_a}} \qquad [4.60a]$$

Thus:

$$r = \left(\frac{3Nv_L^0}{4\pi \mathrm{N_a}}\right)^{1/3} \qquad [4.60b]$$

This can also be written as:

$$N = \frac{4\pi \mathrm{N_a}}{3v_L^0} r^3 \qquad [4.60c]$$

We shall seek to calculate the Reiss function associated with the reaction of condensation [4R.2] in any condition, but where the aggregates are formed in their reference state relation [4.59] can be written:

$$\Delta_{\mathrm{R2},N} G_R = \frac{N}{\mathrm{N_a}} . \Delta_{\mathrm{R2}} g + \sigma_L . \Delta_{\mathrm{R2},N}(A) \qquad [4.61]$$

The variation in area upon the formation of a drop of radius r is:

$$\Delta_{\mathrm{R2},N}(A) = 4\pi r^2 = (4\pi)^{1/3} \left(\frac{3 v_L^0 N}{\mathrm{N_a}} \right)^{2/3} \qquad [4.62]$$

Relation [4.61] then becomes:

$$\Delta_{\mathrm{R2},N} G_R = \frac{N}{\mathrm{N_a}} . \Delta_{\mathrm{R2}} g + \sigma_L (4\pi)^{1/3} \left(\frac{3 v_L^0 N}{\mathrm{N_a}} \right)^{2/3} \qquad [4.63]$$

The differential of this function becomes 0 for the number $N^{\neq}$ of molecules in the nucleus such that:

$$N^{\neq} = -\frac{32\pi}{3} \frac{\left(v_L^0\right)^2 N_a \sigma_L^3}{\left(\Delta_{\mathrm{R2}} g\right)^3} \qquad [4.64a]$$

As we are operating in conditions for which reaction [4R.2] is thermodynamically possible, the term $(\Delta_{R2} g)$ is negative and therefore the corresponding value of $N^{\neq}$ is positive. The corresponding entity is known as the *critical nucleus.*

The Reiss function passes through a maximum whose value is:

$$\Delta_{\mathrm{R2},N} G_R^{\neq} = -\frac{16\pi}{3} \frac{\left(v_L^0\right)^2 \sigma_L^3}{\left(\Delta_{\mathrm{R2}} g\right)^2} = -\frac{N^{\neq} \Delta_{\mathrm{R2}} g}{\mathrm{N_a}} \qquad [4.64b]$$

In addition, the Reiss function takes the value of 0:

$$N^{*} = -\frac{36\pi}{3}\frac{\left(v_L^0\right)^2 N_a}{\left(\Delta_{R2}g\right)^3} \qquad [4.65]$$

Figure 4.4(a) shows the shape of the curve representing function [4.63] with the maximum and the point of intersection of the curve with the abscissa axis.

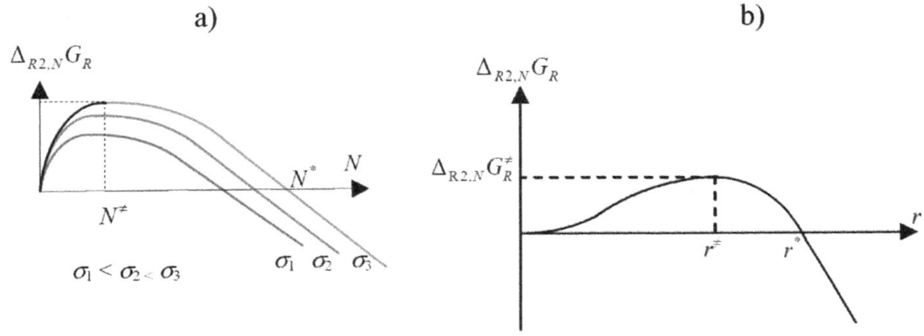

Figure 4.4. *Examples of variation of the Gibbs energy of condensation: a) with the number of construction units in the aggregate; b) with the radius of the nucleus*

If the primary phase is a solution, the reaction of precipitation of the solutes A_i can be written:

$$\sum_i \beta_i A_i = \beta_L L + \sum_j \beta_j A_j \qquad [4R.3]$$

If K_S is the solubility product and Q_S the reaction quotient of the precipitation reaction, we can define the oversaturation as measuring the distance from equilibrium, so:

$$S = \frac{1}{K_S}\frac{\prod_i |A_i|^{\beta_i/\beta_S}}{\prod_j |A_j|^{\beta_j/\beta_S}} = \frac{Q_S}{K_S} \qquad [4.66]$$

In the above expressions, we can reveal the oversaturation by using the relation:

$$\Delta_{\mathrm{R}2} g = -\mathrm{R}T \ln S \qquad [4.67]$$

Hence, relation [4.63] becomes:

$$\Delta_{\mathrm{A},N} G_R = -\frac{N}{\mathrm{N_a}} \mathrm{R}T \ln S + N^{2/3}\left[\sigma_L \left(4\pi N_a\right)^{1/3} \left(3v_L^0\right)^{2/3}\right] \qquad [4.68]$$

We can see that the ordinate of the maximum of the curve in Figure 4.4(a) is smaller when the equilibrium conditions (surface phenomena aside) are surpassed (high value of S), but that this ordinate only becomes 0 in conditions that are infinitely far removed from equilibrium (infinite value of S) – in other words, never.

Below N^*, the associated Reiss function is positive. For larger sizes, it is negative. For this reason, we consider that the nucleus has been formed beyond that size.

If we define the nucleus as being the smallest particle of aggregate which is stable in relation to the precursor A, it will contain N^* construction units, and thus the nucleus will indeed be an aggregate of the size $N = N^*$.

It is easy to verify that $N^{\neq}$ is smaller than N^*.

In the foregoing discussion, we evaluated the Reiss function on the basis of the quantity N. Using relation [4.60b], we can now express the Reiss function and its noteworthy points as a function of the radius of the nucleus. We find the following:

$$\Delta_{\mathrm{R}2,N} G_R = \frac{4\pi r^3}{3v_L^0} \Delta_{\mathrm{R}2} g \; + \; 4\pi r^2 \sigma_L \qquad [4.69]$$

The maximum has the coordinates (critical nucleus):

$$r^{\neq} = -\frac{2\sigma_L v_L^0}{\left(\Delta_{\mathrm{R}2} g\right)^3} \qquad [4.70]$$

and:

$$\Delta_{R2,N}G_R^{\neq} = \frac{16\pi}{3}\frac{\left(v_L^0\right)^2\sigma_L^3}{\left(\Delta_{R2}g\right)^2} = -\frac{4\pi\left(\sigma_L r^{\neq}\right)^2}{3} \quad [4.71]$$

The intersection with the abscissa axis takes place for the maximum radius of the nucleus:

$$r^* = -\frac{3v_L^0\sigma_L}{\left(\Delta_{R2}g\right)} \quad [4.72]$$

Figure 4.4(b) shows the curve satisfying equation [4.69], which is Volmer's curve.

Based on Volmer's curve, we deduce the dimension of the final nucleus, defined as the maximum dimension of the nucleus from which growth proceeds. The division of the heterogeneous reaction into two processes – nucleation and growth – is based on four criteria, the first two thermodynamic in nature, and the last two kinetic in nature:

– criterion no. 1: the two processes are spontaneous;

– criterion no. 2: the effects of surface on the Gibbs energy will be taken into account in nucleation but is negligible in growth;

– criterion no. 3: the solid B already formed has no influence on the mechanism of nucleation, but is involved in that of growth;

– criterion no. 4: the contribution of nucleation to the overall reaction rate (quantity of material transformed per unit time) is negligible in comparison to that of growth.

The first criterion implies that both processes are associated with a negative Reiss–Gibbs energy. If we look again at Volmer's curve (Figure 4.4(b)), the application of this condition at the start of the curve means that nucleation must stop at a value of r greater than r^*.

The second part of the curve, which pertains to growth, occurs beyond the radius $r^{\neq}$. The increase in the dimensions of the condensed phase (growth) will be associated with a negative Gibbs energy $\left(\frac{\mathrm{d}\Delta_{R2,N}G_R}{\mathrm{d}r}<0\right)$.

For the second criterion, beyond a certain point of the downward parabolic branch in Volmer's curve, the surface contribution to the Gibbs energy becomes negligible in comparison to the bulk contribution, and thus we can consider that the molar Gibbs energy no longer varies with the dimension of the grain of the new phase. This is all the more true when the dimension r is significantly greater than r^*. This condition is compatible with the limit of growth set by the previous criterion.

The third criterion, which is a kinetic criterion, states that nucleation must involve the smallest possible quantities of the new phase.

Criterion number four implies the negligibility of the contribution of nucleation in the expression of the reaction rate, and this holds true even more strongly when the new phase is small.

Table 4.2 recaps the conditions we have just examined.

Criterion	Nucleation if	Growth if
1) Spontaneity of the two processes	$r > r^*$	$r > r^{\neq}$
2) Surface contribution	Yes	Not if $r >> r^*$
3) Involvement of B in the mechanism	r as small as possible	
4) Rate of reaction practically given by the growth	r as small as possible	

Table 4.2. *Criteria defining the nucleus of maximum dimensions*

Ultimately, these conditions give rise to a certain contradiction, whereby we need to have as large a transition value r as possible to satisfy criterion 2 and as small as possible (but greater than r^*) for the last two criteria. To resolve this contradiction, though admittedly with only a mediocre degree of precision, we choose the point of ordinate zero ($r = r^*$) as the transition point characterizing the end of the formation of the nucleus, noting that at that point, Volmer's curve is monotonic decreasing, and tends toward a parabolic branch in the direction $\Delta_{R2}g < 0$. The corresponding dimension r^* is the *maximum dimension of the nucleus*.

However, to reach this point, we need to cross the energy barrier, which can only take place because of local fluctuations enabling us to achieve a radius $r \geq r^{\neq}$. Radius $r^{\neq}$, for this reason, is called the *critical radius of the nucleus*, which must not be confused with the maximum radius r^* beyond which specialists in heterogeneous chemical kinetics deem the process of nucleation to be complete and to have been replaced by the process of growth.

NOTE.– In fact, in practice, there is nothing to say that the nucleation should not strictly obey the Volmer curve over time, as it is a growth curve.

4.9.3. *Homogeneous nucleation within a solid phase*

The reasoning process in section 4.9.2 was conducted for a fluid initial phase, but it is worth noting that such a hypothesis is never used in the calculation. This thought process and its results, therefore, apply only to the case of condensation of the defects within a solid phase.

In practice, though, there are a number of *caveats* which must be kept in mind:

– the experimental determination of the promoter(s) is more difficult;

– the energy references need to be carefully monitored;

– the far slower diffusion in solids than in fluid phases will often result in the existence of not-insignificant concentration gradients, meaning that the above reasoning processes can only be applied locally.

4.9.4. *Primary heterogeneous nucleation from a fluid phase*

Experience shows us that, in many cases, the new phase forms from an interface in the initial phase: on the walls of the recipient or the blades of a stirrer in the case of crystallization from a solution, and on the surface of the grains in the case of solid–solid surfaces. As this is an interface with a solid which is not the product of the reaction, it is indeed a primary nucleation, but it is then said to be heterogeneous, because it means that the initial system cannot be treated as a single phase.

We shall now examine the case of the heterogeneous nucleation of a liquid from a vapor phase and the heterogeneous nucleation of a solid from another solid.

4.9.4.1. *Heterogeneous nucleation of a liquid from a vapor*

We shall now discuss the case of the nucleation of a liquid from a vapor onto a solid surface. We suppose that at equilibrium, the nucleus is in the form of a drop with a spherical surface, whose radius is r, resting on the solid substrate (Figure 4.5) and forming an angle θ with that surface, which is the wetting angle of the solid by the liquid.

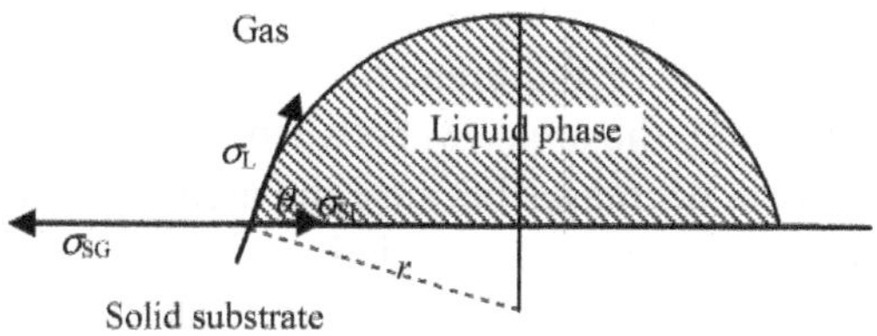

Figure 4.5. *Shape of the nucleus and various interfacial energies in the case of heterogeneous liquid nucleation on a solid substrate*

We shall calculate the variation of the Reiss function upon formation of the nucleus from the gas at a given external pressure and temperature. We choose to use the geometric variables θ and r. The calculation could be performed in the same way if we chose the wetting angle and the number of molecules in the nucleus N as the variables.

We have already calculated (relation [3.29]) the volume of the drop – i.e. the volume of the nucleus – which we can write in the form:

$$V_L = \frac{4\pi r^3}{3} f_{\text{het}} \tag{4.73}$$

The function f_{het}, called the *wetting function*, depends only on θ, and it is a function that varies, as shown by Figure 4.6, between the value 0 if the liquid does not wet the solid at all ($\theta = 0$) and the value 1 obtained when the liquid wets the solid perfectly ($\theta = \pi$).

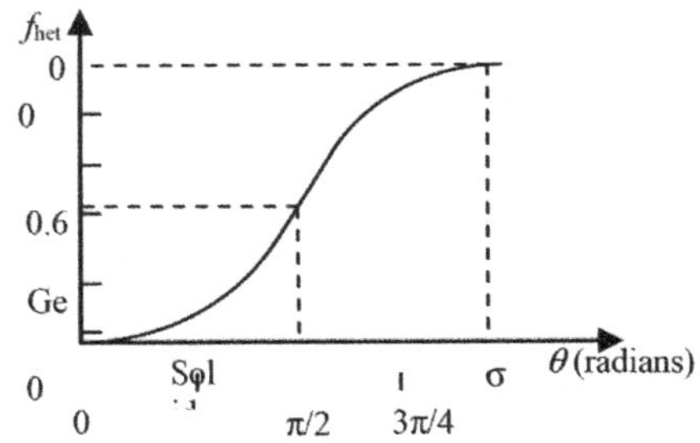

Figure 4.6. *Shape of the curve $f_{het}(\theta)$*

On formation of the nucleus, we see the creation of the area A_{LG} between the liquid and the gas, and of the area A_{SL} between the liquid and the solid support. On the other hand, the area A_{SG} between the gas and the solid support, whose initial value is A_0, decreases because the area A_{SL} is taken away from it. In light of these modifications, the variation of the Reiss function, which always contains a bulk term proportional to the variation of the molar Gibbs energy of the transformation $\Delta_{R2,N}g$ and a surface energy term, is written:

$$\left(\Delta_{R2,N}G_R\right)_{het} = \frac{4\pi r^3}{3v_L^0} f_{het}\Delta_{R2}g + \sigma_{LG}A_{LG} + \left(\sigma_{SL} - \sigma_{SG}\right)A_{SL} \qquad [4.74]$$

Let us evaluate each of the interfacial areas. The area of the interface between the solid and the liquid is that of a disk of radius $r\sin\theta$, so:

$$A_{SL} = \pi r^2 \sin^2\theta \qquad [4.75]$$

The area of the gas/solid interface is given by the initial area of the solid in the absence of the drop A_0 less the area of the previous liquid–solid interface, so:

$$A_{SG} = A_0 - \pi r^2 \sin^2\theta \qquad [4.76]$$

The area of the interface between the gas and the liquid is that of the spherical cap with radius r, centered at O and observed from that point at the solid angle Ω, so:

$$A_{LG} = \Omega r^2 = 2\pi r^2\left(1-\cos\theta\right) \qquad [4.77]$$

The Young–Dupré relation (see relation [3.21]) is written here with our notations:

$$\sigma_{SG} = \sigma_{LG}\cos\theta + \sigma_{SL} \tag{4.78}$$

By introducing the three values of the areas into relation [4.74] and taking account of relation [4.78], we obtain:

$$\begin{aligned}\left(\Delta_{\mathrm{R}2,N}G_R\right)_{\mathrm{het}} &= \frac{4\pi r^3}{3v_L^0} f_{\mathrm{het}}\Delta_{\mathrm{R}2}g \\ &+ 2\pi \mathrm{r}^2\sigma_{LG}\left(1-\cos\theta\right) - \pi \mathrm{r}^2\sigma_{LG}\cos\theta\sin^2\theta\end{aligned} \tag{4.79}$$

By comparing relation [4.79] with relation [4.69], obtained for homogeneous nucleation, we see that it is possible to link the two Reiss functions obtained in heterogeneous and homogeneous conditions by the expression:

$$\left(\Delta_{\mathrm{R}2,N}G_R\right)_{\mathrm{het}} = \left(\Delta_{\mathrm{R}2,N}G_R\right)_{\mathrm{hom}} f_{\mathrm{het}} \tag{4.80}$$

As the function f_{het} is independent of the radius r, the shape of the curve $\left(\Delta_{\mathrm{R}2,N}G_R\right)_{\mathrm{hét}}(r)$ is the same as the curve $\left(\Delta_{\mathrm{R}2,N}G_R\right)_{\mathrm{hom}}(r)$ represented in Figure 4.3(b). The remarkable values of the radii $r^{\neq}$ and r^* are identical and the ordinate of the maximum is given by:

$$\left(\Delta_{\mathrm{R}2,N}G_R^{\neq}\right)_{\mathrm{hét}} = \left(\Delta_{\mathrm{R}2,N}G_R^{\neq}\right)_{\mathrm{hom}} f_{\mathrm{hét}} \tag{4.81}$$

As the value of f_{het} is generally smaller than 1, by comparing equations [4.71] and [4.81], we note that the height of the potential barrier needing to be overcome is less in heterogeneous nucleation than in homogeneous nucleation, which explains why heterogeneous nucleation is easier and therefore more common than homogeneous nucleation. In view of the unavoidable presence of solid dust particles, it is ultimately always heterogeneous nucleation which causes the condensation of a liquid from its vapor.

4.9.4.2. *Heterogeneous nucleation from a solid on another solid*

Consider a system composed of a solid support, a solid nucleus and a fluid phase.

If the interfacial energy between the nucleus and the support is less than the sum of the interfacial energies of the two solids with the fluid phase, it is obvious that the nucleus formed on contact with the support (heterogeneous nucleation) will have a lower energy than that which forms within the fluid (homogeneous nucleation). We have seen the role of the interfacial energy on the Gibbs energy of condensation which will also be lower. To perform a quantitative calculation, it is necessary to establish a set form for the nucleus.

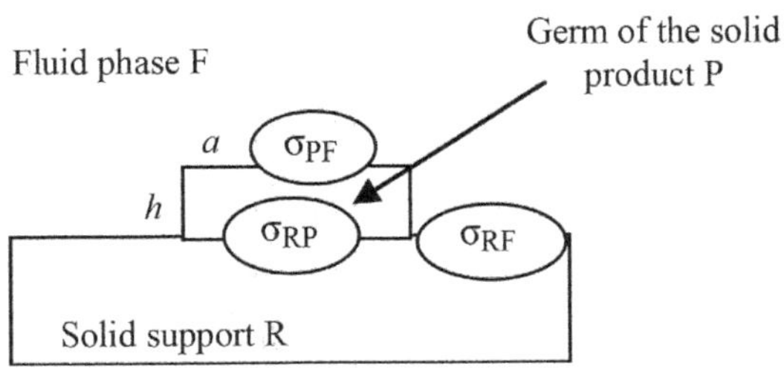

Figure 4.7. *Heterogeneous nucleation of a parallelepipedic solid on a solid support*

We can simplify the problem by looking at the case of a parallelepipedic-shaped nucleus whose interfacial energy does not depend on the face, i.e. on a flat support. Thus, we consider a nucleus with a square surface, with side length a and height h, placed on a flat support (Figure 4.7).

The interfacial energy in the system wherein the agglomerate is not in contact with the support would be:

$$W_{\text{hom}} = \sigma_{\text{RF}} A_0 + \sigma_{\text{PF}}(2a^2 + 4ah) \qquad [4.82]$$

where A_0 is the area of the support.

The juxtaposition of the agglomerate on the support causes the disappearance of two surfaces (RF and PF) and the appearance of an interface RP. Thus, the interfacial energy becomes:

$$W_{\text{het}} = \sigma_{\text{RF}} A_0 + \sigma_{\text{PF}}(2a^2 + 4ah) + (\sigma_{\text{RP}} - \sigma_{\text{RF}} - \sigma_{\text{PF}})a^2 \qquad [4.83]$$

The volume V_P of the nucleus is fixed by the number of construction units of the agglomerate:

$$V_P = a^2 h = \frac{N}{N_a} v_P^0 \quad [4.84]$$

The shape of the nucleus (ratio a/h) is given by the minimal surface energy. However, this energy is minimal when the differential in relation to a of the surface energy is zero, such that:

$$\frac{dW_{het}}{da} = 0 \quad [4.85]$$

Consider:

$$\frac{2V_P}{a^3} = 1 + \frac{\sigma_{RP} - \sigma_{RF}}{\sigma_{PF}} = 1 - \frac{\sigma_{RF} - \sigma_{RP}}{\sigma_{PF}} \quad [4.86]$$

Let us set:

$$p = \frac{\sigma_{RF} - \sigma_{RP}}{\sigma_{PF}} \quad [4.87]$$

Relation [4.86] becomes:

$$\frac{1-p}{2} = \frac{V_P}{a_3} = \frac{h}{a} \quad [4.88]$$

Examine the ranges of value of p:

– if the interface energy created by juxtaposition of the two solids is greater than the sum of those taken away, there will be no sticking, and relation [4.88] must not be applied. This situation corresponds to:

$$\sigma_{RP} - \sigma_{RF} - \sigma_{PF} > 0 \text{ so } p < -1$$

– if, however, sticking is favored ($p > 0$), the ratio h/a tends toward zero: the solid P spreads as much as possible over the support R. On the other hand, a value of p greater than 1 leads to an abnormal situation ($h < 0$).

Thus, we define the value p, called the *spreading parameter*, of the interfacial system (R, P, F) as follows:

– if the spreading coefficient h_t (see relation [3.8]) is positive – i.e. if:

$$\sigma_{RP} - \sigma_{RF} - \sigma_{PF} \geq 0 \quad [4.89a]$$

$$\text{then } p = -1 \quad [4.89b]$$

– if the spreading coefficient is positive, and if:

$$0 \geq \sigma_{RP} - \sigma_{RF} - \sigma_{PF} \geq 2\sigma_{PF} \quad [4.90a]$$

$$\text{then } p = \frac{\sigma_{RF} - \sigma_{RP}}{\sigma_{PF}} \quad [4.90b]$$

– if the spreading coefficient is positive, and if:

$$0 \geq \sigma_{RP} - \sigma_{RF} - \sigma_{PF} \leq 2\sigma_{PF} \quad [4.91a]$$

$$\text{then } p = +1 \quad [4.91b]$$

This definition also has the advantage of being consistent with the formalism applied to liquids, where p is the cosine of the wetting angle.

For the morphology at equilibrium, by substituting relations [4.88] and [4.90] back into relation [4.83], we obtain:

$$W_{het} = \sigma_{RF}.A_0 + 3\sigma_{PF}.a^2.(1 - p) \quad [4.92]$$

We can now evaluate the Gibbs energy associated with the formation of a mole of nuclei on the surface of the solid support, with each nucleus containing N construction units:

– the system in question contains the solid support, in an indifferent quantity but sufficient so that the adsorption of the nuclei is not sterically inhibited, the initial solution and the nuclei formed:

- the chemical term remains: $(N/N_a)\Delta_{R2}g$;

- the interfacial term, which is obtained by subtracting the energy in the initial state (support plus solution with no nucleus) $\sigma_{RF}.A_0$ from the value in the final state, given by relation [4.92], modified to apply to one mole of nucleus:

$$W_{het} = 3\sigma_{PF}.a^2.(1 - p) \quad [4.93]$$

Using relation [4.84], we can write:

$$\frac{N}{N_a} = \frac{a^3}{v_P^0}\frac{1-p}{2} \quad [4.94]$$

We can see that this value becomes 0 when $p = 1$: in the case of *total wetting*, there is no longer a nucleation barrier.

In light of relation [4.94], the sum of the two terms gives us:

$$\Delta_{R2,N}(G_R)_{het} = \frac{a^3}{v_P^0}\frac{(1\text{-p})}{2}\Delta_{R2}g + 3\sigma_{PF}.a^2(1 - p) \quad [4.95]$$

Let us define an equivalent radius and an apparent interfacial tension by the relations:

$$a^3 = (A_{RP})^{3/2} = \frac{4\pi}{3}r^3 \quad [4.96a]$$

and:

$$\sigma_S = \left(\frac{3}{4\pi}\right)^{1/3}\sigma_{PF} \quad [4.96b]$$

After reorganizing the terms, and when these values are fed back into the previous relation, we find:

$$\Delta_{R2,N}(G_R)_{het} = \frac{4\pi r^3}{3v_P^0}\frac{(1\text{-}p)}{2}\Delta_{R2}g + 4\pi\sigma_S r^2\frac{(1 - p)}{2} \quad [4.97]$$

We recognize an expression similar to that which was established for homogeneous primary nucleation [4.69], with each term multiplied by a value that depends only on the geometry and the spreading parameter. We shall denote this value as f_{het}:

$$\Delta_{\text{R2},N}\left(G_R\right)_{\text{het}} = \frac{4\pi r^3}{3v_{\text{P}}^0} f_{\text{het}}\ \Delta_{\text{R}2}g + 4\pi\sigma_{\text{S}} r^2 f_{\text{het}} \qquad [4.98]$$

where $f_{\text{het}} = \dfrac{1-p}{2}$ [4.99]

The function f_{het} will be called the *wetting function.*

In view of the definition of p, f_{het} varies between 0 (*total wetting*, with the disappearance of the interfacial term) and 1 (*no wetting*). In the latter case, heterogeneous nucleation has the same energy as homogeneous nucleation.

We saw earlier that in the case of a drop in form of a spherical cap, we are led to the same result, with the definition [4.73] of f_{het} in accordance with:

$$f_{het} = \frac{2-3\cos\theta+\cos^3\theta}{4} = \frac{(2+p)(1-p)^2}{4} \qquad [4.100]$$

This definition is consistent with the equation:

$$p = \cos\theta \qquad [4.101]$$

Generally speaking, we consider that for each form, it is possible to define a function f_{het}, ranging from 0 for $p = 1$ (total wetting) to 1 for $p = -1$ (no wetting), such that the Reiss potential for nucleus formation is of the form given in equation [4.98].

Hence, for each form, we can define the function f_{het} on the basis of the equilibrium dimensions of the nucleus, as we did on the basis of equation [4.85], of the term σ_{S} and of the equivalent radius, as we did with relations [4.96].

Of course, in all cases, the curves of the Reiss function as a function of the equivalent radius have the same form as shown in Figure 4.4(b), with a maximum and an intersection of the abscissa axes given by the same expressions. The concepts of *critical nucleus* and *nucleus of maximum dimensions* are thus preserved.

5

Capillary Tubes and Thin Films

In the previous chapter, we examined phases with small dimensions in all directions. This chapter is devoted to the study of phases which exhibit small dimensions in only one or two directions. Such phases will either be capillary tubes (cylindrical or flat) or thin liquid films.

5.1. Behavior of a liquid in a capillary space

We shall examine the thermodynamic properties of a liquid phase placed in a capillary space (in a cylindrical tube or between two parallel plates that are close together) in the presence of a gaseous phase.

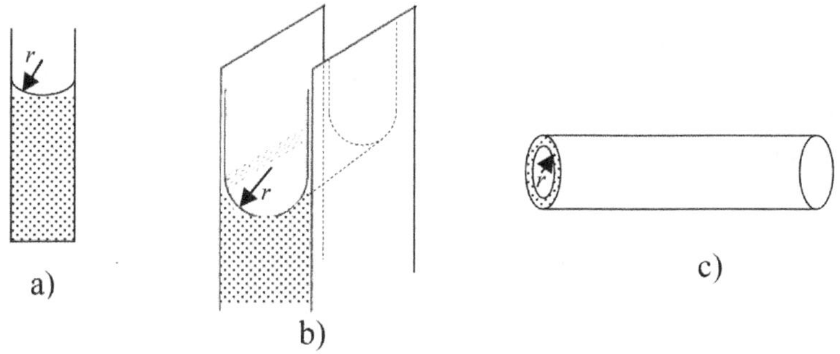

Figure 5.1. *a) Spherical meniscus in a cylindrical tube; b) cylindrical meniscus between two plates; c) cylindrical meniscus on the walls of a cylinder*

This liquid phase can be separated from the gaseous phase either by a spherical meniscus, as will be the case in a cylindrical capillary tube

(Figure 5.1(a)) or by a cylindrical meniscus which may develop between two planes (Figure 5.1(b)) or in a cylindrical tube (Figure 5.2(c)).

We studied the thermodynamic properties linked to the spherical meniscus in the previous chapter (see sections 4.1 to 4.8). We shall now examine the thermodynamic properties linked to the cylindrical meniscus.

5.2. Thermodynamics of the cylindrical meniscus

The thermodynamics of the cylindrical meniscus has an effect, as with the case with the spherical meniscus, on the variance of the equilibria of state change, and also, because of the presence of a non-limited dimension of a phase, on the extent of the phase by the phenomenon of capillary ascension.

5.2.1. *Laplace's law for the cylindrical meniscus*

Consider the expression of Laplace's law in the vicinity of a cylindrical curve of radius r which is of the form:

$$P^{(\text{int})} = P^{(\text{ext})} + \frac{\sigma}{r} \tag{5.1}$$

It is easy, with a cylindrical meniscus, to carry out a study similar to that performed for the bubble or the spherical meniscus in sections 4.1 to 4.8. In the case of a liquid meniscus, the inside (as understood in the sense of Laplace's law) is the gas and the outside is the liquid. Thus, we can deduce the saturating vapor pressure above the cylindrical meniscus of radius r which, based on relation [4.33], will be:

$$P^{(\text{vap})}(r) = P^{(vap)}(\infty)\exp\left(-\frac{\sigma\, v^{0(\text{liq})}}{\text{R}Tr}\right) \tag{5.2}$$

This is Kelvin's law for the cylindrical meniscus. The radius r_K at equilibrium at the saturating vapor pressure $P^{(\text{vap})}(r_k)$ is sometimes called the Kelvin radius for the cylinders for that pressure.

Obviously, in the same way, we can study other properties – e.g. the melting of a small solid cylinder in a capillary tube or between two parallel planes. With the hypothesis of a liquid cylinder with the same height as the solid cylinder from which it comes, we find the relation:

$$T_f^{\infty} - T_f = \frac{v^{0(\text{sol})}}{\Delta_f S}\left[\sigma^{(\text{sol})} - \sigma^{(\text{liq})}\left(\frac{v^{0(\text{liq})}}{v^{0(\text{sol})}}\right)\right]\frac{1}{r^{(\text{sol})}} \qquad [5.3]$$

The difference between the melting points of the bulk phase and the capillary phase is qualitatively identical to that given by equation [2.39] in the case of droplets, meaning that this difference is inversely proportional to the radius of the cylinder.

5.2.2. *Capillary ascension*

In this chapter, we consider a liquid phase of small dimensions in contact with a gaseous bulk phase.

5.2.2.1. *Contact of a liquid with a wall*

Consider a liquid in a recipient whose walls are supposed to be flat. With those walls, the liquid forms a contact angle θ, determined by Young's law. This causes deformation of the surface of the liquid in the vicinity of the wall (Figure 5.2).

We want to calculate the profile of that deformation. It is clear that the mean radius of curvature at a point on the surface separating the liquid and the gas above it is equal to the radius of curvature at the point with coordinates (z, x) in the plane xOz, meaning that at every point, the surface behaves like a cylindrical meniscus with radius r. Therefore we can apply Laplace's law [2.1], written in the new context, which gives, in the vicinity of the surface separating the two phases, the pressure in the liquid as a function of the pressure in the gaseous phase in the form:

$$P^{(\text{liq})} = P^{(\text{ext})} - \frac{\sigma}{r} \qquad [5.4]$$

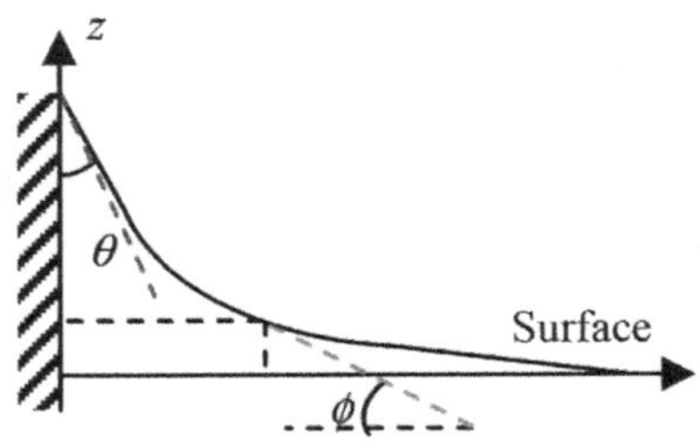

Figure 5.2. *Contact of a liquid with a vertical wall*

If we assume the liquid incompressible with is density ρ, the distribution of pressure in the liquid gives us:

$$P^{(\text{liq})} = P^{(\text{ext})} - \rho g z \quad [5.5]$$

From relations [5.4] and [5.5], we deduce the equality:

$$\rho g z = \frac{\sigma}{R} \quad [5.6]$$

If ϕ denotes the angle of the Ox axis with the tangent to the curve of abscissa x, and if s is the curvilinear abscissa along the curve in the "upward" direction, the local curvature of a planar curve is given by:

$$\frac{1}{r} = \frac{\mathrm{d}\varphi}{\mathrm{d}s} \quad [5.7]$$

The curvilinear abscissa is such that:

$$\mathrm{d}s = \sqrt{\mathrm{d}x^2 + \mathrm{d}z^2} \quad [5.8]$$

Although the tangent of the angle ϕ is the derivative of the function $z(x)$, so:

$$\tan\varphi = \frac{\mathrm{d}z}{\mathrm{d}x} \quad [5.9]$$

By combining expressions [5.6], [5.7], [5.8] and [5.9], we find:

$$\rho g z = \sigma \frac{\mathrm{d}\varphi}{\mathrm{d}s} = \frac{\sigma \,\mathrm{d}\varphi}{\mathrm{d}x\sqrt{1+\tan^2\varphi}} = \frac{\sigma \cos\varphi \tan\varphi \,\mathrm{d}\varphi}{\mathrm{d}z} \qquad [5.10]$$

which can also be written in the form:

$$\rho g z dz = \sigma \sin\varphi \,\mathrm{d}\varphi \qquad [5.11]$$

By integration on both sides of the above equation, we obtain:

$$\frac{1}{2}\rho g z^2 = -\sigma \cos\varphi + Cte \qquad [5.12]$$

To determine the constant, we note that if $z = 0$, meaning that at the level of the horizontal plane of the liquid, far from the wall, then the angle φ is zero and $\cos\varphi = 1$. By feeding this property back into expression [5.12], we obtain:

$$\frac{1}{2}\rho g z^2 = \sigma(1-\cos\phi) = 2\sigma \sin^2\frac{\varphi}{2} \qquad [5.13]$$

Such is the equation of the profile of deformation of the meniscus.

Let us now define the capillary length l_c by the expression:

$$l_c = \sqrt{\frac{\sigma}{\rho g}} \qquad [5.14]$$

By feeding this formula back into relation [5.12], the new equation for the meniscus is:

$$z = 2 l_c \sin\frac{\varphi}{2} \qquad [5.15]$$

We can see that the deformation along the wall is of the order of magnitude of the capillary length. When $\theta = 0$, then $\varphi = 0$ and thus the maximum height which it is possible to achieve is:

$$z_{\max} = 2 l_c \qquad [5.16]$$

Thus, for water, the capillary length is approximately 1.5 mm, which gives us a maximum height of a little more than 3 mm.

5.2.2.2. *Jurin's law*

If we immerse a capillary tube in a liquid, under the influence of the surface in the vicinity of the wall, the level of the liquid in the tube is different to its level in the recipient (Figure 5.3(a)). A height h separates the two levels.

We suppose that the capillary effects are predominant in the tube, meaning that the radius of the tube r_c is smaller than the capillary length l_c defined by relation [5.13]. We can then consider that the pressure of the liquid is the same across the whole of the surface of the meniscus. This surface is assumed to be spherical (this hypothesis is not necessary, but it is very complicated to calculate h without it) with radius r.

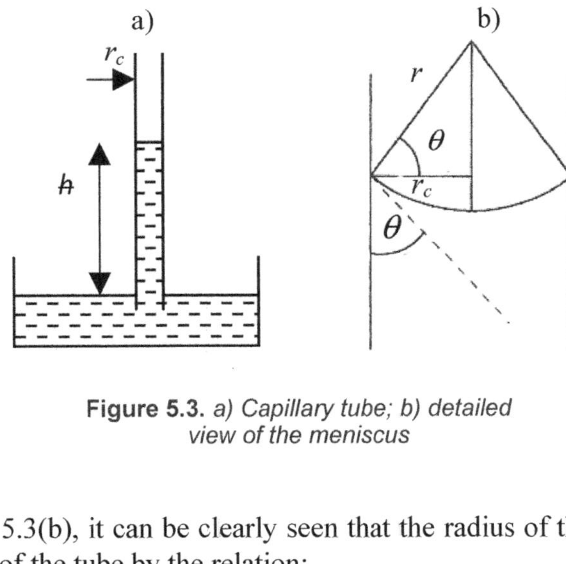

Figure 5.3. *a) Capillary tube; b) detailed view of the meniscus*

In Figure 5.3(b), it can be clearly seen that the radius of the cap is linked to the radius of the tube by the relation:

$$r = \frac{r_c}{\cos\theta} \qquad [5.17]$$

where θ is the contact angle between the liquid and the wall, or the *wetting angle*.

The pressure of the liquid at the meniscus, therefore, by virtue of Laplace's law for a spherical meniscus, is:

$$P^{(\text{liq})} = P^{(\text{ext})} - 2\frac{\sigma}{r} \quad [5.18]$$

This pressure is also that of the gas in the tank minus the weight of the liquid between the two levels, such that, if ρ is the density of the liquid and g is the acceleration due to gravity:

$$P^{(\text{liq})} = P^{(\text{ext})} - \rho g z \quad [5.19]$$

By comparing relations [5.18] and [5.19], we immediately obtain:

$$h = \frac{2\sigma\cos\theta}{\rho g r_c} = \frac{2\cos\theta}{r_c} l_c^2 \quad [5.20]$$

This law is *Jurin's law*, which thus specifies that the height h is inversely proportional to the radius of the tube.

For instance, in the case of water in a glass capillary ($\theta = 0$) for a tube 1 cm in radius, the height reaches 14 mm. For a radius of 1 μm, the height would then be 14 m, which leads to a negative pressure, because atmospheric pressure is equivalent to a column of water of height 10.33 m. It is unsurprising that the liquid exerts a force of "attraction" on the walls, as does a metal under the force of traction.

In Figures 5.3, it was assumed that the liquid wetted the tube ($\theta < \pi/2$). If the liquid does not wet the tube ($\theta > \pi/2$), the opposite effects are observed: the level of the liquid in the tube is lower than in the recipient (Figures 5.4(a) and 5.4(b)). Such is the case, for example, of mercury with glass, where we have $\theta \cong 140°$.

Figure 5.4. *Capillary drop: a) capillary tube; b) contact of liquid with the vertical wall*

These properties of Laplace's and Jurin's laws are applied for determining the size of the pores of a solid material using a mercury porosimeter.

Imagine a porous solid, placed in a bath of mercury, on which a pressure $P(r_c)$ is exerted. Initially, imagine that all the pores in the solid are cylinders with the radius r_c. We would then see a sudden increase in the volume of mercury which would be absorbed by those pores as soon as the pressure reaches the value which compensates for the surface tension, given by:

$$P(r_c)=\frac{2\sigma\cos\theta}{r_c} \qquad [5.21]$$

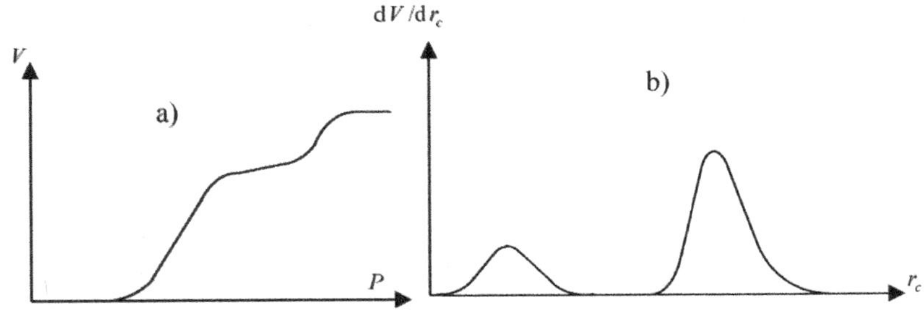

Figure 5.5. *a) Curve showing the filling as a function of the pressure; b) distribution of the pores in terms of volume*

In reality, we actually see a more gradual increase, which corresponds to the progressive filling of increasingly small pores (Figure 5.5(a)). By measuring the volume absorbed as a function of the pressure applied, we are able to work out the size distribution of the pores of the solid under examination.

Let $V(r_c)$ denote the total volume of the pores of the material whose radius is less than r_c, and V_0 the total volume of the pores in the material. We measure the difference $V=V_0-V(r_c)$ as a function of the pressure. Based on that difference, we can write:

$$\frac{\mathrm{d}V(r_c)}{\mathrm{d}r_c}=\frac{\mathrm{d}V}{\mathrm{d}P}\frac{\mathrm{d}P}{\mathrm{d}r_c}=-\frac{P}{r_c}\frac{\mathrm{d}V}{\mathrm{d}P} \qquad [5.22]$$

We deduce:

$$\frac{\mathrm{d}V}{\mathrm{d}r_c} = \frac{P}{r_c}\frac{\mathrm{d}\left(V_0 - V(r)\right)}{\mathrm{d}P} \qquad [5.23]$$

Thus, by knowing the dependence of the volume on P, we are able to calculate the distribution of the pores in the volume (Figure 5.5(b)). The pressures applied are high (up to 1000 bars, which corresponds to radii of 700 nm) and the solid must obviously be capable of resisting crushing at such pressures.

Practical determination of the pore radius distribution is performed, using the experimental curve (Figure 5.5(a)), by dividing the abscissa axis into equal intervals and choosing, at the middle of each interval, a mean radius and attributing to that mean radius the ratio of the difference in volumes to the difference in radii on either side of that mean value.

NOTE.– In reality, the pores are not cylindrical and the mercury input radius is the radius of aperture of the pore; on the other hand, the volume measured is the true volume of the pore corresponding to that inlet radius. The pores, therefore, are considered to be cylinders (Figure 5.6) having a height h such that:

$$h = \frac{V_p}{\pi r_c^2} \qquad [5.24]$$

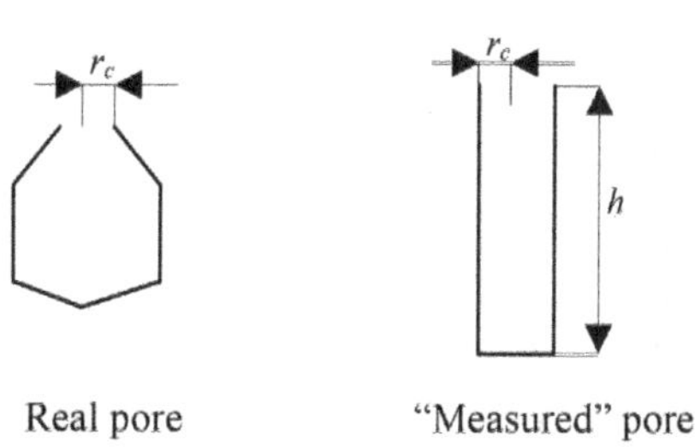

Figure 5.6. *Real pore and fictitious cylindrical measured pore*

5.2.3. *Capillary condensation*

In view of Kelvin's law [4.31], the saturating vapor pressure of a liquid in a space of dimensions of the order of magnitude of the capillary length l_c is less than the saturating vapor pressure of the same liquid in a space of large dimensions. This results in condensation in the capillary space at vapor pressures less than the "normal" vapor pressure of the liquid at equilibrium. Thus, the liquid may appear in the capillary space, although in a larger volume the vapor would remain dry. This is the phenomenon of *capillary condensation*.

5.2.3.1. *Capillary condensation in a cylindrical medium*

Consider a cylindrical pore with radius r_c. Above that pore, the pressure of a substance is $P^{(\text{vap})}$ (Figure 5.7(a)).

Various domains of remarkable values of the pressure $P^{(\text{vap})}$ need to be considered:

– if the pressure $P^{(\text{vap})}$ is less than the saturating vapor pressure in the pore ($P^{(\text{vap})} < P^{0(\text{vap})}(r_K)$), r_K is the Kelvin radius defined by:

$$r_K = \frac{r_c}{\cos\theta} \quad [5.25]$$

The pore is empty and the vapor is dry.

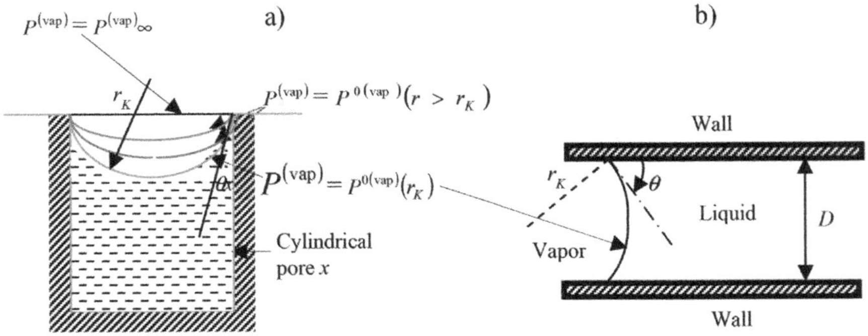

Figure 5.7. *Filling of a pore: a) in a cylinder; b) between two plates*

– if the pressure $P^{(\text{vap})}$ is equal to the saturating vapor pressure in the pore ($P^{(\text{vap})} = P^{0(\text{vap})}(r_K)$), the liquid condenses in the pore and fills it until the level of the liquid reaches the edge of the tube, respecting the wetting angle θ. This pressure is given by relation [4.33], because the meniscus with the smallest radius of curvature liable to form is a spherical cap whose radius is equal to the Kelvin radius r_K defined by relation [5.23], and satisfying:

$$r_K = \frac{\sigma\, v^{0(\text{liq})}}{2\text{R}T \ln \dfrac{P^{0(vap)}(\infty)}{P^{0(\text{vap})}(r_K)}} \qquad [5.26]$$

– if the vapor pressure lies between Kelvin's value and the "normal" saturating vapor pressure – i.e. if we have $P^{0(\text{vap})}(r_K) < P^{(\text{vap})} < P^{0(\text{vap})}(\infty)$ – then the vapor will condense in the tube with meniscus radii greater than r_K. As the vapor pressure increases, the radius of curvature of the meniscus also increases, and the tube fills (Figure 5.7(a)) until an infinite radius is reached. In other words, the result is a full pore with a horizontal interface.

5.2.3.2. *Capillary condensation between two flat plates*

Consider two parallel plates, situated a distance D apart from each other, which is of the order of magnitude of the length of the Kelvin radius (Figure 5.7(b)). The meniscus likely to form between the two plates is a cylindrical meniscus. The Kelvin saturating vapor pressure for the cylinders is given, if we ignore the compressibility of the liquid, by relation [5.2]. In view of that relation, the Kelvin radius of the cylindrical meniscus is:

$$r_K = \frac{\sigma\, v^{0(\text{liq})}}{\text{R}T \ln \dfrac{P^{0(vap)}(\infty)}{P^{0(\text{vap})}(r_K)}} \qquad [5.27]$$

where $r_K < 0$.

For a given vapor pressure, condensation will occur between the two plates if the distance D is such that:

$$D < D_k = -2r_K \cos\theta \qquad [5.28]$$

For distances larger than D_K, the vapor remains dry; for a distance equal to D_K, the meniscus forms, and the space between the plates is filled with liquid in such a way that the wetting angle of the solid by the liquid is respected. With distances greater than D_K, the radius of curvature of the cylindrical meniscus increases with that distance D until it is practically infinite, thus leading to a quasi-planar surface.

5.3. Modeling the interactions between two surfaces of an insulating material

It may seem surprising that we should discuss the modeling of this phenomenon here, in a chapter which is devoted to thin films, when it might have been expected to be in Chapter 1, when we were looking at liquid surfaces. This is due to the fact that, when two materials are in contact with one another, between the two surfaces there is always a third substance, in the form of a thin thread – generally air. The system is then formed of two interfaces between different materials.

Thus, let us begin by considering two surfaces of two materials (one of them solid, liquid or gaseous), 1 and 2 (which may be identical). Between two elements (atoms or molecules), A and B, belonging to each of the two materials, there are forces of interaction created by a potential $\varepsilon_{\mathrm{A/B}}$:

$$\vec{f}_{\mathrm{A/B}} = -\operatorname{grad} \varepsilon_{\mathrm{A/B}} \qquad [5.29]$$

At a sufficiently small distance between the surfaces of the two materials, that potential $\varepsilon_{\mathrm{A/B}}$ begins to no longer be negligible in comparison to the other forms of energy – kinetic, potential, etc.

If we place the surfaces of these two materials face to face and consider an elementary volume $\mathrm{d}\tau$ of material 1 at a distance d from a molecule of material 2 (see Figure 5.8(a)), where v_1^0 and v_2^0 respectively denote the molecular volumes of materials 1 and 2, the energy of interaction per unit volume between those two elements would be:

$$\mathrm{d}\varepsilon_{\mathrm{surf}\,1/2} = \frac{\varepsilon_{\mathrm{A/B}}(r)}{v_1^0 v_2^0}\mathrm{d}\tau \qquad [5.30]$$

Hence, by integrating for the whole of the volume of material 1, we find the energy of the molecule of material 2 placed at a distance d from material 1 to be:

$$\varepsilon_{\text{surf}\,1/2} = \iiint_{r>d} \frac{\varepsilon_{\text{A/B}}(r)}{v_1^0 v_2^0} \mathrm{d}\tau \qquad [5.31]$$

By integrating on the volume of material 2, we obtain the surface energy of material 1 placed at a distance e from material 2 (Figure 5.8(b)):

$$W_{1/2}(e) = \int_e^{\infty} \varepsilon_{\text{surf}\,1/2}(d)\,\mathrm{d}r \qquad [5.32]$$

This energy is expressed per unit surface. It is linked to a surface force which is exerted on the walls of the surfaces in accordance with:

$$F_{1/2} = -S\frac{\mathrm{d}W_{1/2}(e)}{\mathrm{d}e} \qquad [5.33]$$

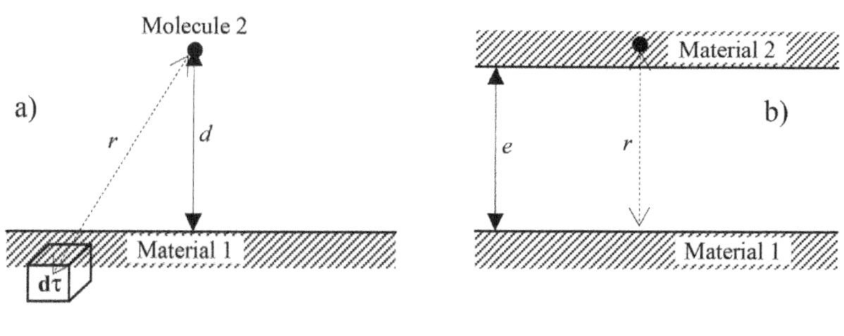

Figure 5.8. *Interactions between a surface and: a) a molecule; b) another surface*

The most common form of interaction between molecules is given by the van der Waals forces. We know that if we discount the forces of repulsion in $1/r^{12}$ in comparison to that of attraction in $1/r^6$, the Lennard–Jones attraction potential between two molecules A and B situated a distance r apart is given by the relation:

$$\varepsilon_{\text{A/B}} = -\frac{m}{r^6} \qquad [5.34]$$

The coefficient m is the sum of the three contributions of orientation, induction and dispersion, and depends on the permanent electrical moments and particularly on the polarizabilities of the molecules of material 1 and material 2.

Let us apply the integration [5.31] on the basis of Figure 5.9. Operating in spherical coordinates, if Ω denotes the solid angle from which a spherical cap in material 1 is observed from the molecule of 2, the elementary volume of material 1 is given by:

$$\mathrm{d}\tau = \Omega r^2 \,\mathrm{d}r = 2\pi r^2 \left(1-\cos\theta\right)\mathrm{d}r \qquad [5.35]$$

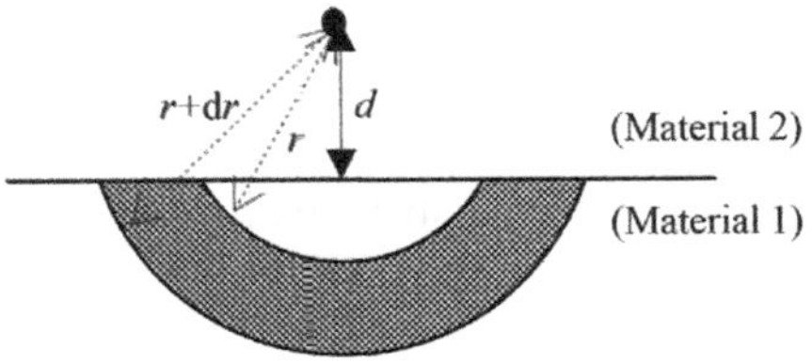

Figure 5.9. *Integration of the energy of a molecule of (2) in relation to the surface of (1) in spherical symmetry*

By feeding this back into relation [5.31], for the energy per unit volume of material 2 placed at a distance d from a surface of material 1, we obtain:

$$\varepsilon_{\text{surf}\,1/2} = \frac{1}{v_1^0 v_2^0}\int_r^{\infty} 2\pi r^2 \left(1-\cos\theta\right)\left(-\frac{m}{r^6}\right)\mathrm{d}r \qquad [5.36]$$

This gives us:

$$\varepsilon_{\text{surf}\,1/2} = \frac{-\pi m}{6 v_1^0 v_2^0 r^3} \qquad [5.37]$$

Now let us apply integration [5.32]. We immediately obtain the following for the energy per unit surface of a surface of material 1 placed at a distance e from the material 2:

$$W_{1/2} = \frac{-\pi m}{12 e^2 v_1^0 v_2^0} = \frac{-A_{\mathrm{H}}^{(12)}}{12\pi e^2} \qquad [5.38]$$

The value $A_{\mathrm{H}}^{(1/2)}$ is called that *Hamaker constant* for materials 1 and 2. It is expressed in joules and is defined by:

$$A_{\mathrm{H}}^{(1/2)} = \frac{\pi^2 m}{v_1^0 v_2^0} \qquad [5.39]$$

As it is the effect of dispersion which is often greatest in the different contributions to the van der Waals interaction, the most important values in m are the polarizabilities. Thus, we can write that the Hamaker constant is of the order of magnitude of:

$$A_{\mathrm{H}}^{(1/2)} \approx \frac{\alpha_1 \alpha_2}{v_1^0 v_2^0} \qquad [5.40]$$

where α_1 and α_2 are the polarizabilities of phases 1 and 2.

The values of the Hamaker constant, in fact, vary fairly little from one material to another. Table 5.1 gives a few typical values for a number of families of materials.

Materials	A
Low-energy surface (organic solid)	10^{-21} J
High-energy surface (ceramics)	$\approx 10^{-18}$ J
Vapor phase	negligible

Table 5.1. *Noteworthy values of the Hamaker constant*

The fact that the Hamaker constant is negligible for vapor or for air renders the properties of surfaces in contact with air or a vapor absolutely identical to those obtained in contact with a vacuum.

NOTE.– The calculation we have performed is applicable to insulating materials, but it cannot be applied to electrical conductors, because of the mobility of the electrons, which produces forces that are very different to the van der Waals forces.

Now consider the stack of three materials, such as a film of liquid L between two materials 1 and 2, as illustrated by Figure 5.10. For this case,

Lifschitz performed a similar calculation to the one performed above. The expression of the energy of interaction between materials 1 and 2 through the liquid is again approximately given by:

$$W_{1/2} = \frac{-\pi m}{12e^2 v_1^0 v_2^0} = \frac{-A_H^{(1/L/2)}}{12\pi e^2} \quad [5.41]$$

The Hamaker constant then depends on the dielectric constants of the three materials.

Hamaker constants are generally tabulated for interfaces composed of the same material separated by a vacuum or a solvent. To calculate the constant between two different materials, we can use the approximation of the geometric mean, in the form:

$$A_H^{(1/L/2)} \cong \sqrt{A_H^{(1/L/1)} A_H^{(2/L/2)}} \quad [5.42]$$

The study of the interactions between surfaces brings us back to the notion of surface energy (which is the surface tension for liquids). This is defined as being the energy needed to create two surfaces by separating a given volume of material into two. We can write, if a_0 is the size of one molecule:

$$2\sigma = W(\infty) - W(a_0) \quad [5.43]$$

If $A_H^{(1/V)}$ is the Hamaker constant of the interface between material 1 and the vacuum, we deduce a new expression for the surface tension:

$$\sigma = \frac{A_H^{(1/V)}}{24\pi a_0^2} \quad [5.44]$$

With this relation, it is possible to establish the tables of the Hamaker constant.

We can also look again at the energy of adhesion if we write, for a medium 1, in view of relation [5.43]:

$$2\sigma_1 = W_{11} \quad [5.45]$$

From this, we deduce the interfacial tension between media 1 and 2:

$$\sigma_{12} = \frac{1}{2}(W_{11} + W_{22}) - W_{12} = \sigma_1 + \sigma_2 - W_{12} \qquad [5.46]$$

Then, for the energy of adhesion, we find Dupré's expression [3.76].

Relations [5.44] and [5.46] are indeed satisfied for apolar molecules. However, they are less exact in the case of polar molecules, for which the approximation of the exclusive contribution of the forces of dispersion to the van der Waals forces is less accurate.

5.4. Thin liquid films

We shall now look at a new category of small-dimension phases, with thin liquid films on the surface of a liquid. Such films are a few nanometers in thickness. It is important not to confuse a thin liquid film with an interface between two liquid phases, even though the dimensions may be similar. Indeed, the phase of which a thin film is composed of is of a different nature to the phase which supports it, and unlike interfaces, constitutes an independent phase. However, these films possess specific properties because of their low thickness, which means that a certain amount of interaction can take place between the two bulk phases surrounding the films.

5.4.1. *Disjunction pressure*

Consider a thin liquid film L, of thickness e and area A, trapped between two phases 1 and 2 (Figure 5.10), at least one of which is condensed. For instance, phase 1 might be liquid and phase 2 might be air.

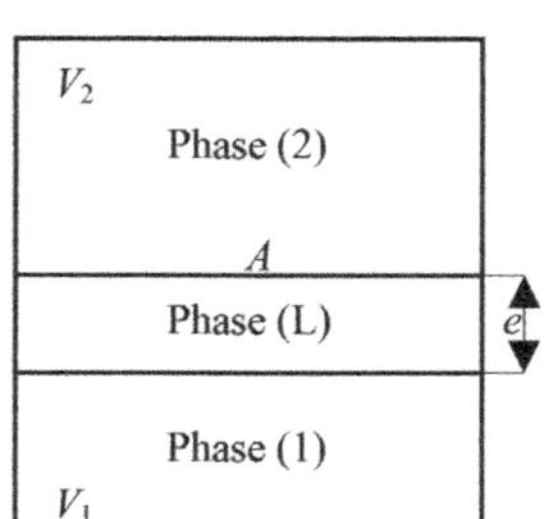

Figure 5.10. *Liquid film between two bulk phases*

We make a small variation in the thickness e (e.g. an increase by de) by adding liquid. The volumes of the phases 1 and 2 are sufficiently great to be assumed to remain constant in the elementary transformation.

We can now show that the pressure $P^{(L)}$ in the thin film is different from the external pressure $P^{(ext)}$ in phases 1 and 2 in the vicinity of the interfaces.

The total Helmholtz energy of the system is written:

$$F = U - TS + A\left[\sigma_{1/L} + \sigma_{2/L} + W_{12}(e)\right] \qquad [5.47]$$

$W_{12}(e)$ is the energy of interaction between the surfaces of phases 1 and 2 through the liquid. In the elementary transformation whereby the thickness varies by de, the variation in Helmholtz energy, at constant temperature and volumes V_1 and V_2, where V is the volume of the film, will be:

$$\mathrm{d}F = P^{(ext)}\,\mathrm{d}V - P^{(L)}\,\mathrm{d}V + A\frac{\mathrm{d}W_{12}}{\mathrm{d}e}\mathrm{d}e \qquad [5.48]$$

In view of the fact that the volume of the film is given by the product Ae and that the area A is kept constant, this variation in Helmholtz energy can be written as:

$$\mathrm{d}F = AP^{(ext)}\,\mathrm{d}e - AP^{(L)}\,\mathrm{d}e + A\frac{\mathrm{d}W_{12}}{\mathrm{d}e}\mathrm{d}e \qquad [5.49]$$

We define the *disjunction pressure* $\Pi_d(e)$ of the film by the difference:

$$\Pi_d(e) = P^{(ext)} - P^{(L)} \qquad [5.50]$$

The equilibrium condition of the film is expressed by a minimum value of the Helmholtz energy, so $\mathrm{d}F = 0$. By combining this condition with expressions [5.49] and [5.50], we obtain:

$$\Pi_d(e) = -\frac{\mathrm{d}W_{12}}{\mathrm{d}e} \qquad [5.51]$$

Two scenarios may arise:

– if $\Pi_d(e) > 0$, the two surfaces tend to repel one another, meaning that the film tends to become thicker by the addition of matter. In this case, the liquid wets phases 1 and 2 perfectly (contact angles of 0);

– if $\Pi_d(e) < 0$, the two surfaces tend to come closer together, and the liquid tends to be ejected. We then say that it wets phases 1 and 2 imperfectly (contact angles between 0 and 90°).

Of course, if the liquid does not wet the two phases 1 and 2, no film can form and the problem no longer arises.

In the case of van der Waals interactions, by applying relation [5.41], the disjunction pressure becomes:

$$\Pi_e(e) = -\frac{A_{\mathrm{H}}^{(1/\mathrm{L}/2)}}{6\pi e^3} \quad [5.52]$$

Thus, the disjunction pressure is a force tending to separate the two interfaces of a thin film with the phases surrounding it.

5.4.2. *Formation of a film by condensation*

Consider a solid surface, covered with a thin film of liquid in contact with its own vapor at a pressure $P^{(\mathrm{vap})}$ less than its saturating vapor pressure $P^{0(\mathrm{vap})}$. In view of the existence of a disjunction pressure, the vapor pressure at equilibrium with the thin liquid film must be different from the saturating vapor pressure $P^{0(\mathrm{vap})}$. The pressure in the liquid film is imposed by the relation:

$$P^{(\mathrm{liq})} = P_{film}^{0(\mathrm{vap})} - \Pi_d(e) \quad [5.53]$$

The molar Gibbs energy of a liquid varies with the pressure, in accordance with:

$$\frac{dg^{0(\mathrm{liq})}}{dP} = v^{0(\mathrm{liq})} \quad [5.54]$$

If g^0_{sat} is the Gibbs energy of the liquid (or the vapor) at normal equilibrium between the two phases, the Gibbs energy of the liquid at the pressure $P^{(\text{L})}$ would be:

$$g^{0(\text{liq})} = g^0_{\text{sat}} + v^{0(\text{liq})}\left(P^{(\text{liq})} - P^{0(\text{vap})}\right) \quad [5.55]$$

The Gibbs energy of the vapor at pressure $P^{(\text{V})}$ would be:

$$g^{(\text{V})} = g^0_{\text{sat}} + RT\ln\frac{P^{(\text{V})}}{P^{0(\text{vap})}} \quad [5.56]$$

Equality of the chemical potentials arises at a pressure of equilibrium between the liquid and the vapor $P^{(\text{V})} = P^{0(\text{film})}$, such that:

$$\frac{\text{R}T}{v_M^{(\text{L})}}\ln\frac{P^{0(vap)}_{film}}{P^{0(\text{vap})}} = \left(P^{(liq)} - P^{0(\text{vap})}\right) \quad [5.57]$$

In light of relation [5.53], this expression can also be written:

$$\frac{\text{R}T}{v_M^{(\text{L})}}\ln\frac{P^{0(\text{vap})}_{\text{film}}}{P^{0(\text{vap})}} = \left(P^{0(\text{vap})}_{\text{film}} - P^{0(\text{vap})} - \Pi_e\right) \quad [5.58]$$

However, if we calculate the term $\text{R}T/v^{0(\text{liq})}$, it is such that we can accept the approximation:

$$\frac{\text{R}T}{v^{0(\text{liq})}}\ln\frac{P^{0(\text{vap})}_{\text{film}}}{P^{0(\text{vap})}} >> \left(P^{0(\text{vap})}_{\text{film}} - P^{0(\text{vap})}\right) \quad [5.59]$$

Relation [5.58] is simplified to give us:

$$\frac{\text{R}T}{v^{0(\text{liq})}}\ln\frac{P^{0(\text{vap})}_{\text{film}}}{P^{0(\text{vap})}} = -\Pi_e \quad [5.60]$$

From this, we deduce the pressure at equilibrium between the vapor and the film:

$$P^{0(\text{film})} = P^{0(\text{vap})} \exp - \frac{v^{0(\text{liq})} \Pi_e}{\text{R}T} \qquad [5.61]$$

We can see that it is possible to condense liquid at a pressure lower than the saturating vapor pressure, if the disjunction pressure is positive.

If we consider that van der Waals interactions are responsible for the disjunction pressure, using relation [5.52], the vapor pressure at equilibrium with the film would be:

$$P_{\text{film}}^{0(\text{vap})} = P^{0(\text{vap})} \exp \frac{A_{\text{H}}^{(\text{liq/V})} v^{0(\text{liq})}}{6\pi \text{R}Te^3} \qquad [5.62]$$

Thus, we can see, in the case that the Hamaker coefficient between the liquid and its vapor – i.e. between the liquid and a vacuum – is negative, or if the liquid wets the solid, then if we start with a very low value and increase the vapor pressure, there comes a moment when a thin liquid film of the compound is deposited.

In this case, we speak of *pre-wetting* rather than of wetting, because the phase obtained is stable in the film state but unstable as a bulk, because the vapor pressure remains less than the saturating vapor pressure.

5.4.3. *Ascension of a liquid along a wall*

We shall now show that when a liquid is contained in a recipient, a film of that liquid, of height h and thickness e, which is a function of h, is thermodynamically stable along the walls of the recipient.

Thus, consider a recipient containing a liquid (Figure 5.11). A film of that liquid is supposed to be formed on the walls. We shall now examine whether this is true. The pressure within the film, according to relation [5.50], is:

$$P^{(\text{L})}(h) = P_{atm} - \Pi_d(e(h)) \qquad [5.63]$$

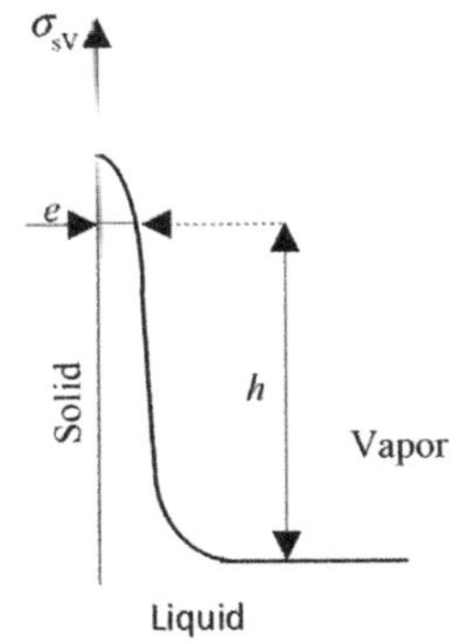

Figure 5.11. *Ascension of a liquid along a wall*

We can see that this column of liquid has a certain weight, and therefore that, if ρ is the density of the liquid, we also have mechanical equilibrium, which yields:

$$P^{(L)}(h) = P_{atm} - \rho gh \tag{5.64}$$

By comparing equations [5.63] and [5.64], we find:

$$\Pi_d(e(h)) = -\frac{dW}{de} = \rho gh \tag{5.65}$$

This equation gives us the thickness of the film at the height h.

This film, then, is stable, but exists only if the disjunction pressure is positive – i.e. if the liquid perfectly wets the solid forming the recipient.

We shall evaluate the thickness and the height of that film at thermodynamic equilibrium.

The energy (i.e. the Helmholtz energy) of the liquid per unit length along the wall is given by:

$$F = \int_0^h [\rho gez + \sigma_{SL} + \sigma_{LV} + W(e) - \sigma_{SV}]\mathrm{d}z \tag{5.66}$$

If we reveal the spreading coefficient defined by relation [3.81], by:

$$h_t = \sigma_{SV} - \sigma_{SL} - \sigma_{LV} = \sigma_{LV}(\cos\theta - 1) \qquad [5.67]$$

the equilibrium condition of the film in relation to its dimensions is:

$$\frac{\mathrm{d}F}{\mathrm{d}h} = 0 \qquad [5.68]$$

which gives us the equilibrium for a maximum height h_{max} and a thickness e_c:

$$\rho g h_{max} e_c - h_t + W(e) = 0 \qquad [5.69]$$

In light of relation [5.59], this condition becomes:

$$\Pi_d . e_c - h_t + W = 0 \qquad [5.70]$$

Thus, this relation defines the thickness at equilibrium at the maximum height of the film.

For instance, in the context of van der Waals interactions, in view of relations [5.41] and [5.51], equation [5.70] is written:

$$h_t - \frac{A_H^{(S/L/V)}}{2\pi e_c^2} + e_c \frac{A_H^{(S/L/V)}}{6\pi e_c^3} = 0 \qquad [5.71]$$

By solving this equation, we find the value of e_c, and by substituting that value back into equation [5.69], we obtain the following for e_c and the maximum height:

$$\begin{cases} e_c = \sqrt{\dfrac{-A_H^{(S/L/V)}}{4\pi h_t}} \\ h_{max} = \dfrac{-A_H^{(S/L/V)}}{6\pi\rho g e_c^3} \end{cases} \qquad [5.72]$$

For typical values of the spreading coefficient and the Hamaker constant, the minimum thickness e_c is around a nanometer – i.e. approximately equal to the size of a molecule – and the film spreads over the walls until it forms a monomolecular layer.

This film is the explanation for the Rollin films obtained, in particular, with helium below a certain temperature. If we evaluate the height of that liquid in the case of helium with a Hamaker constant of around 10^{-19} J, and accept the hypothesis that the thickness of the film is 1 nm, the height may be up to 500 m. Thus, we can understand why, in a recipient of reasonable dimensions, liquid helium escapes from the recipient by flowing above the walls. This phenomenon is less marked in other liquids, because it is counteracted by viscosity, which is zero in the case of superfluid helium.

5.4.4. *Minimum spreading thickness*

Now let us consider a puddle of liquid deposited on a solid support which it wets perfectly (see Figure 5.12). The thickness of that puddle is supposed to be uniform, and the amount of liquid and hence the volume of the puddle are also taken to be constant. We shall now evaluate the minimum thickness of the puddle at equilibrium if there are no walls to limit its spreading.

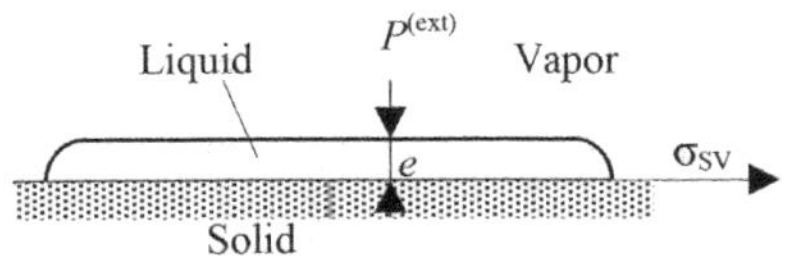

Figure 5.12. *Maximum spreading of a drop on a solid support*

The Helmholtz energy of the puddle considered to be a thin film of negligible mass is as follows (the volume of gas above the puddle and the support is assumed to be constant):

$$F = U - TS + AW - h_t A \qquad [5.73]$$

Its differential in a transformation would be:

$$\mathrm{d}F = T\,\mathrm{d}S - P^{(L)}\,\mathrm{d}V^{(L)} + A\frac{\mathrm{d}W}{\mathrm{d}e}\mathrm{d}e + W\,\mathrm{d}A - h_t\,\mathrm{d}A \qquad [5.74]$$

If that transformation is the spreading of the puddle, its volume remains constant, having the value:

$$V^{(\mathrm{L})} = Ae \qquad [5.75]$$

which we write as:

$$dV^{(\mathrm{L})} = 0 \qquad [5.76]$$

From this, we deduce:

$$A\,\mathrm{d}e = -e\,\mathrm{d}A \qquad [5.77]$$

Hence, at constant temperature and volume of liquid T and $V^{(\mathrm{L})}$, the differential of the Helmholtz energy (relation [5.74]) becomes:

$$\mathrm{d}F = \left(-\frac{\mathrm{d}W}{\mathrm{d}e} + W - h_t\right)\mathrm{d}A \qquad [5.78]$$

The equilibrium condition obtained for the minimum of the function F will occur with a thickness e^* obeying the equation:

$$-\frac{\mathrm{d}W}{\mathrm{d}e^*} + W - h_t = 0 \qquad [5.79]$$

Here we see the same equation as relation [5.70]; thus, the thickness at equilibrium is the same as that of the vertical film on the walls: $e^* = e_c$. Hence, this thickness is roughly that of a monomolecular layer. This low thickness is the reason for the interference taints often observed on such films.

6

Physical Adsorption of Gases by Solids

Any time a gas is in the presence of a solid surface, we know that a certain number of molecules of that gas will attach to the surface of the solid. This is the phenomenon of *adsorption*. The reverse process, by which molecules detach from the surface and enter the gaseous medium, is known as *desorption*.

6.1. Shapes of the isotherms of physical adsorption found experimentally

The main manifestation of the phenomenon of adsorption is shown by the curves known as isotherms of adsorption, which represent the quantity adsorbed as a function of the pressure of the gas at a specific temperature. The aim of thermodynamic studies is to attach physical meaning and mathematical expressions to these curves. With that goal in mind, Brunauer [BRU 40] established a classification of the experimental curves found into five types (Figure 6.1). These curves show the quantity of gas adsorbed n_a as a function of the ratio P/P^0, where P^0 denotes the pressure of liquefaction of the gas at the same temperature.

Type I, known as a *Langmuir isotherm*, is found with non-porous or microporous solids, where the pore diameter is less than 250 nm. The horizontal part of the curve corresponds to the saturation of the surface of the solid with a monolayer of molecules of gas. For all other types, adsorption takes place with several layers of gas affixed to the surface.

Type II isotherms are essentially found with macroporous solids – i.e. solids whose pore diameters are greater than 2000 nm.

Type III isotherms are found more infrequently, and are a derivative of type II, as described above.

Type IV isotherms are found with porous solids where the pore diameters are between 250 and 2000 nm, sometimes referred to as *mesopores*. These isotherms exhibit a plateau of saturation.

Type V isotherms are less common, and have a shape derived from that of type IV isotherms.

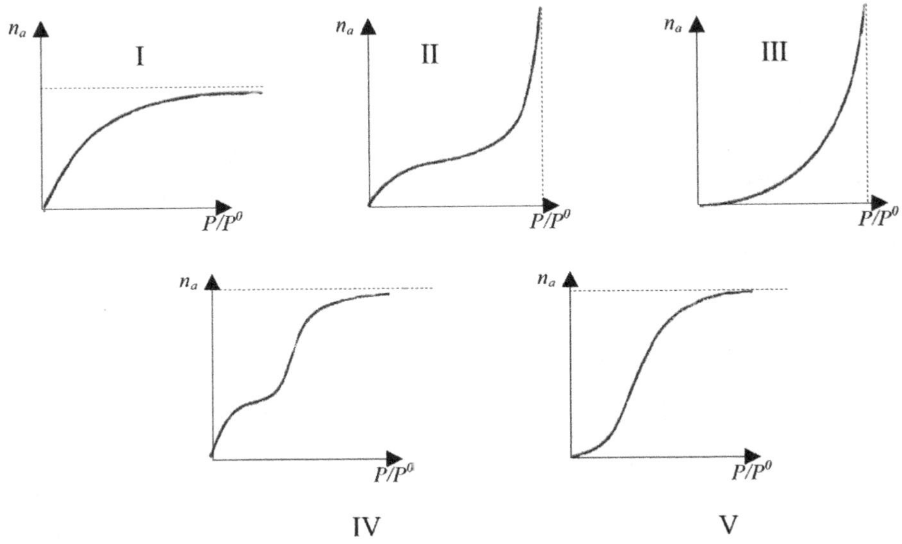

Figure 6.1. *The five types of isotherms of physical adsorption [BRU 40]*

6.2. Potential energy of a gaseous molecule in the presence of the surface of a solid

The essential value in determining the behavior of a molecule of gas in the vicinity of the surface of a solid is the potential energy of that molecule under the influence of the solid. Lennard-Jones recognized that the forces exerted from a surface on a molecule of gas are of different natures depending on whether the solid is an electrical insulator or an electron conductor such as a metal.

6.2.1. *Adsorbent insulating solid*

If the solid is an insulator, the force exerted between the solid and the gaseous molecule G is the resultant of the van der Waals forces exerted between the gaseous molecule G and each of the molecules of the solid. We know that these van der Waals forces derive from a potential which, between two molecules A and B, includes two terms – a term of attraction and a term of repulsion – which lead to a distance of equilibrium r_0 between the two molecules A and B.

Let us first examine the term of attraction, which is written in the form:

$$\varepsilon_a^{(\mathrm{AB})} = -\frac{m_0}{r^6} = 2\varepsilon_0\left(\frac{r_0}{r}\right)^6 \qquad [6.1]$$

For the whole of the solid, we find the sum of each of the terms of attraction between the molecule G and each molecule of the solid. If we let N be the number of molecules of solid per unit volume, the overall potential of attraction of the molecule will be formulated as:

$$\varepsilon_a^{(\mathrm{G})} = \int_z^{\infty} \varepsilon_a^{(AB)} N \,\mathrm{d}V \qquad [6.2]$$

We express the elementary volume composed of a spherical cap of the solid centered at the molecule (Figure 6.2(a)) in the form:

$$\mathrm{d}V = S\,\mathrm{d}r = 2\pi r^2\left(1-\cos\theta\right)\mathrm{d}r = 2\pi r^2\left(1-\frac{z}{r}\right)\mathrm{d}r \qquad [6.3]$$

If we combine expressions [6.1], [6.2] and [6.3], the attraction term is shown to be:

$$\varepsilon_a^{(\mathrm{G})} = \frac{\pi N \varepsilon_0 r_0^6}{3z^3} \qquad [6.4]$$

We can see that an intermolecular attraction potential which is inversely proportional to the distance between molecules to the power of 6 results in an attraction potential between the molecule G and the solid which is inversely proportional to the distance z to the power of 3.

Let us now turn our attention to the van der Waals repulsion potential. Between two molecules A and B, it is expressed in the form:

$$\varepsilon_r^{(AB)} = \frac{l_0}{r^{12}} = -\varepsilon_0\left(\frac{r_0}{r}\right)^{12} \qquad [6.5]$$

By performing the same integration as for the attraction potential, we find the repulsion potential between the molecule of gas and the solid:

$$\varepsilon_r^{(G)} = \frac{\pi N\varepsilon_0 r_0^{12}}{45z^9} \qquad [6.6]$$

Thus, an intermolecular repulsion potential inversely proportional to the distance between molecules, to the power of 12, leads to a repulsion potential between the molecule G and the solid which is inversely proportional to the distance z, to the power of 9.

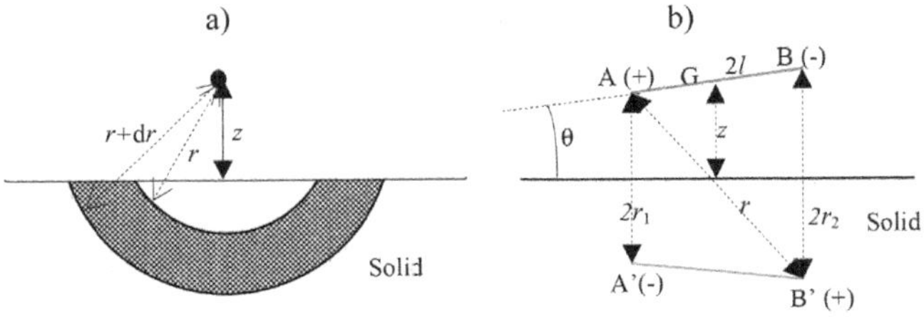

Figure 6.2. *Calculation of the energy of a molecule of gas in the presence of a solid: a) insulating solid; b) conductive solid*

By combining the two contributions, [6.4] and [6.6], the potential energy of the molecule G in the presence of the solid is written as:

$$\varepsilon^{(G)} = \frac{\pi N\varepsilon_0 r_0^3}{3}\left[\left(\frac{r_0}{z}\right)^3 - \frac{1}{15}\left(\frac{r_0}{z}\right)^9\right] \qquad [6.7]$$

The curve given the potential energy as a function of the distance z does have a minimum (Figure 6.3), corresponding to the equilibrium distance z_0, which is obtained by zeroing the differential of the function [6.7] in relation to z, giving us:

$$z_0 = \frac{r_0}{5^{1/6}} \qquad [6.8]$$

The term r_0 is a distance characteristic of the couple formed by the molecule G and the molecule which constitutes the solid. Note that the distance z_0 is smaller than the distance r_0. At equilibrium, the molecule of G is nearer to the surface than an isolated molecule of the solid would be.

6.2.2. *Electronically-conductive adsorbent solid*

Lennard-Jones noted that the potential given by relation [6.7] was not applicable when the solid is an electronic conductor. The mobility of the free electrons under the influence of the distribution of charges in the adsorbed molecule led Lennard-Jones to put forward a model where the electrons in the metal form the instantaneous electrical image of the charge distribution, variable over time, of the molecule. This model ignores the relaxation time of the electrons in the solid – in other words, the assumption is made that those electrons have infinite mobility.

We shall now calculate this Lennard-Jones potential in a case where the molecule of gas has a permanent electrical moment: $\mu = 2ql$.

At a given time, the molecule G, which is at a distance z from the solid, has an electrical moment which forms an angle α with the surface of the solid, assumed to be limitless (Figure 6.2(b)).

The forces of attraction are exerted between the charges q, placed at A and A' on the one hand, and also at B and B' on the other. The intensities of those forces are:

$$F_{AA'} = -\frac{q^2}{(2r_1)^2} \qquad [6.9a]$$

and:

$$F_{BB'} = -\frac{q^2}{(2r_2)^2} \qquad [6.9b]$$

The attraction potential between the molecule and the surface would be:

$$\varepsilon_a^{(G)} = -\int_z^{\infty}\left[\frac{q^2}{\left(2r_1^2\right)^2}+\frac{q^2}{\left(2r_2^2\right)^2}\right]\mathrm{d}z \qquad [6.10]$$

In Figure 6.2(b), we use geometric logic to calculate:

$$r_1 = z - l\sin\theta \qquad [6.11a]$$

and:

$$r_2 = z + l\sin\theta \qquad [6.11b]$$

By feeding these expressions back into relation [6.10] and calculating the integral, we find the following for the attraction potential between the molecule and the solid:

$$\varepsilon_a^{(G)} = -\frac{q^2}{4z}\left(\frac{1}{z-l\sin\theta}+\frac{1}{z+l\sin\theta}\right) \qquad [6.12]$$

The forces of repulsion are exerted between the charges placed at points A and B' on the one hand, and A' and B on the other. Their intensities are, in both cases:

$$F_{AB'} = F_{A'B} = \frac{q^2}{r^2} \qquad [6.13]$$

Thus, the repulsion potential can be calculated by the integral:

$$\varepsilon_r^{(G)} = \frac{q^2}{2}\int_z^{\infty}\frac{1}{r^2}\mathrm{d}z \qquad [6.14]$$

In Figure 6.2(b), it is easy to calculate:

$$r = 2\sqrt{z^2 + l^2\sin^2\theta} \qquad [6.15]$$

If we adopt the hypothesis that $z^2 >> l^2 \sin^2\theta$, the repulsion potential is then written:

$$\varepsilon_r^{(G)} = \frac{q^2}{2z}\left(1 + \frac{l^2\cos^2\theta}{3z^2}\right) \quad [6.16]$$

Hence, in view of the two potentials of attraction (relation [6.12]) and repulsion (relation [6.16]), the total interaction potential between the molecule G and the electronically conductive solid becomes:

$$\varepsilon^{(G)} = -\frac{q^2}{2z} - \frac{q^2}{6z^3}l^2\cos^2\theta + \frac{q^2}{2z}\frac{1}{1-\frac{l^2\sin^2\theta}{z^2}} \quad [6.17]$$

This can alternatively be written:

$$\varepsilon^{(G)} = -\frac{q^2}{6z^3}l^2\cos^2\theta + \frac{q^2}{2z^3}l^2\sin^2\theta \quad [6.18]$$

The dipole G is mobile over time, and the potential varies arbitrarily, so we use an angle α that is the average of the mean values of the square cosine and the square sine. We can calculate the mean value $\overline{\cos^2\theta}$ by the expression:

$$\overline{\cos^2\theta} = \frac{\int_0^\pi \cos^2\theta\,\mathrm{d}\theta}{\int_0^\pi \mathrm{d}\theta} = \frac{\int_0^\pi \frac{1-\cos\theta}{2}\mathrm{d}\theta}{\int_0^\pi \mathrm{d}\theta} = \frac{1}{2} \quad [6.19]$$

Similarly, for $\overline{\sin^2\theta}$, we have:

$$\overline{\sin^2\theta} = 1 - \overline{\cos^2\theta} = \frac{1}{2} \quad [6.20]$$

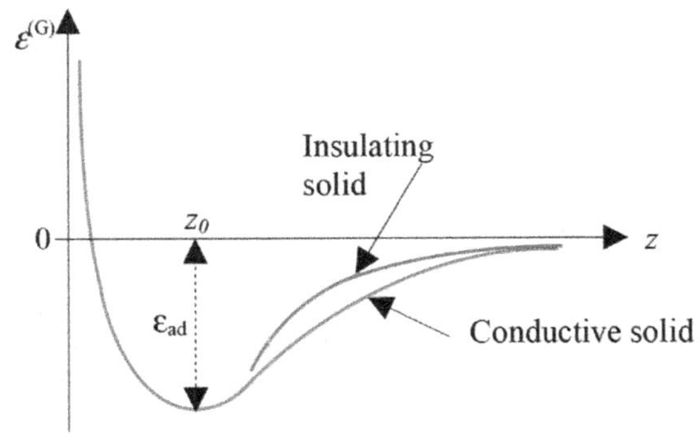

Figure 6.3. *Potential for interaction of a molecule of gas in the vicinity of a solid*

In light of the expression of the electrical moment:

$$\mu = 2ql \quad [6.21]$$

the interaction potential becomes:

$$\varepsilon^{(G)} = -\frac{\mu^2}{12z^3} \quad [6.22]$$

If the molecule does not have a permanent dipole moment, we can use the same relation provided we replace the permanent moment with the mean dipole moment due to the internal fluctuations in the molecule, and the interaction potential becomes:

$$\varepsilon^{(G)} = -\frac{\overline{\mu^2}}{12z^3} \quad [6.23]$$

Relations [6.22] and [6.23] do not give us the equilibrium position for the adsorbed molecule (no minimum on the corresponding curve in Figure 6.3), because of the approximation of z^2 and l^2 which we made in order to bring the calculation to fruition. The potential calculated is thus only valid for a satisfactory distance between the molecule and the surface of the solid. That distance does not include the minimum which would tell us the position at equilibrium.

6.3. Thermodynamic models for physical adsorption

In order to study the thermodynamics of the process of physical adsorption, it is necessary to imagine a model which defines a state for the adsorbed molecule. This model needs to take two important experimental results into account:

– the variance of the system is 2, meaning that the quantity adsorbed at equilibrium is a function of the gas pressure and the temperature;

– the phenomenon of adsorption is exothermic, which means that an increase in temperature decreases the quantity adsorbed at equilibrium.

Two states have been envisaged for the adsorbed species. One led to the development of Hill's model, and the other to Hill and Everett's model.

6.3.1. *Hill's model*

In this model, the adsorption layer – i.e. the ensemble of adsorbate and adsorbed material – is assimilated to a solution of the adsorbed gas G and the adsorbate solid S. Thus, we can apply the general properties of solutions. The possible variables are: the pressure, assimilated to the partial pressure of G in the case of a pure gas (the partial pressure of the solid S is negligible); the temperature; the adsorbed quantity of G in solution; the quantity of S in solution. If the quantity of adsorbed gas, $n_{\langle\langle G\rangle\rangle}$, is denoted by n_a, then the number of independent components is 2 (G and S), the number of external intensive parameters is 2, the pressure and temperature and the number of phases is also 2: the solid solution and the gas. Thus, the Gibbs variance is 2. Hence, we can define:

– isotherms giving the quantity of gas adsorbed $n_a = f(P)$ at constant temperature;

– isobars given $n_a = f(T)$ at constant pressure;

– isosteres $P = f(T)$ at constant adsorbed quantity n_a.

6.3.1.1. *General equation for equilibrium in Hill's model*

With regard to the adsorbed phase (solution), we work with a constant quantity of adsorbate, $dn_{\langle\langle S\rangle\rangle}=0$. The variation of the chemical potential of the gas $\langle\langle G\rangle\rangle$ in the adsorbed phase is:

$$\mathrm{d}\mu_{\langle\langle G\rangle\rangle}=-\overline{S_{\langle\langle G\rangle\rangle}}\,\mathrm{d}T+\overline{V_{\langle\langle G\rangle\rangle}}\,\mathrm{d}P+\left(\frac{\partial\mu_{\langle\langle G\rangle\rangle}}{\partial n_a}\right)_{P,T,n_S^{(A)}}\mathrm{d}n_a \qquad [6.24]$$

The variation of the chemical potential of the adsorbed material G in the gaseous phase with a single component is:

$$\mathrm{d}g^0_{\{G\}}=-s^0_{\{G\}}\,\mathrm{d}T+v^0_{\{G\}}\,\mathrm{d}P \qquad [6.25]$$

Equilibrium of the adsorbed material G between the two phases requires that the variations of its chemical potential be equal between the two phases in the case of any disturbance – i.e. that:

$$\mathrm{d}\mu_{\langle\langle G\rangle\rangle}=\mathrm{d}g^0_{\{G\}} \qquad [6.26]$$

In view of expressions [6.24] and [6.25], this equality is expressed by:

$$\left(s^0_{\{G\}}-\overline{S_{\langle\langle G\rangle\rangle}}\right)\mathrm{d}T+\left(v^0_{\{G\}}-\overline{V_{\langle\langle M\rangle\rangle}}\right)\mathrm{d}P+\left(\frac{\partial\mu_{\langle\langle G\rangle\rangle}}{\partial n_a}\right)_{P,T,n_{\langle\langle S\rangle\rangle}}\mathrm{d}n_a=0 \qquad [6.27]$$

From this general expression, giving all of the points of equilibrium, we shall deduce the equations of the different curves: isosteres, isotherms and isobars.

6.3.1.2. *Equation of the isostere in Hill's model*

To find the equation of the isostere, we apply the general equation [6.27] at a constant quantity of adsorbed G, such that: $dn_a=0$. This relation becomes:

$$\left(\frac{dP}{dT}\right)_{n_a}=\frac{s^0_{\{G\}}-\overline{S_{\langle\langle G\rangle\rangle}}}{v^0_{\{G\}}-\overline{V_{\langle\langle G\rangle\rangle}}} \qquad [6.28]$$

If we assume the molar volume of the gas to be much greater than the partial molar volume of the component G in the adsorbed layer:

$$v^0_{\{G\}} >> \overline{V_{\langle\langle G\rangle\rangle}} \qquad [6.29]$$

and that the gas is perfect:

$$v^0_{\{G\}} = \frac{RT}{P} \qquad [6.30]$$

we can then write for the difference in volume:

$$v^{0)}_{\{G\}} - \overline{V_{\langle\langle G\rangle\rangle}} \cong v^{0)}_{\{G\}} \cong \frac{RT}{P} \qquad [6.31]$$

Furthermore, by definition of the molar enthalpy, we can write in each of the two phases:

$$\mu_{\langle\langle G\rangle\rangle} = \overline{H_{\langle\langle G\rangle\rangle}} - T\overline{S_{\langle\langle G\rangle\rangle}} \qquad [6.32]$$

and

$$g^0_{\{G\}} = h^0_{\{G\}} - Ts^0_{\{G\}} \qquad [6.33]$$

At equilibrium, the chemical potentials in the two phases are equal, and thus:

$$\mu_{\langle\langle G\rangle\rangle} = g^0_{\{G\}} \qquad [6.34]$$

We define the isosteric heat of adsorption (see section 6.3) on the basis of the difference between the entropies (first equation [6.35]), and by combining relations [6.32], [6.33] and [6.34], we obtain the difference between the enthalpies, and we have:

$$\frac{q_{\text{isost}}}{T} = s^0_{\{G\}} - \overline{S_{\langle\langle G\rangle\rangle}} = \frac{h^0_{\{G\}} - \overline{H_{\langle\langle G\rangle\rangle}}}{T} \qquad [6.35]$$

Let the isosteric heat of adsorption be:

$$q_{\text{isost}} = h^0_{\{G\}} - \overline{H_{\langle\langle G\rangle\rangle}} \qquad [6.36]$$

By feeding expressions [6.31], [6.35] and [6.36] into relation [6.28] and ignoring the variations in isosteric heat with differing temperature, we obtain the equation of the isostere:

$$\left(\frac{d \ln P}{dT}\right)_{n_{\langle\langle M\rangle\rangle}} = \frac{q_{\text{isost}}}{\mathrm{R}T^2} \qquad [6.37]$$

As we saw earlier, this equation makes the assumption that the gas is perfect – an approximation which is often perfectly reasonable in the phenomenon of adsorption, particularly when we are not too near to the saturation close to the condensation of the gas.

6.3.1.3. *Equation of the isotherm in Hill's model*

From the general equation [6.27], we can deduce the equation for the isotherm by making $\mathrm{d}T = 0$, which gives us:

$$\left(v^0_{\{G\}} - \overline{V_{\langle\langle G\rangle\rangle}}\right)\mathrm{d}P + \left(\frac{\partial \mu_{\langle\langle G\rangle\rangle}}{\partial n_{\mathrm{a}}}\right)_{P,T,n_{\langle\langle S\rangle\rangle}} \mathrm{d}n_{\mathrm{a}} = 0 \qquad [6.38]$$

Using the same hypotheses about the volumes which led us to approximation [6.31], the equation of the isotherm is written:

$$\left(d \ln P\right)_T = \frac{1}{\mathrm{R}T}\left(\frac{\partial \mu_{\langle\langle G\rangle\rangle}}{\partial n_{\mathrm{a}}}\right)_{P,T,n_{\langle\langle S\rangle\rangle}} \mathrm{d}n_{\mathrm{a}} \qquad [6.39]$$

The model of a solution used to express the chemical potential as a function of the composition at a chosen temperature will give the expression of its differential in relation to that same composition and will enable us to make relation [6.39] more precise.

6.3.1.4. *Equation of the isobar in Hill's model*

From the general equation [6.27], we deduce the equation of the isobar by making dP = 0, which gives us:

$$\left(s^{0)}_{\{G\}} - \overline{S_{\langle(G)\rangle}}\right)\mathrm{d}T + \left(\frac{\partial \mu_{\langle(G)\rangle}}{\partial n_a}\right)_{P,T,n_{\langle(S)\rangle}} \mathrm{d}n_a = 0 \qquad [6.40]$$

In light of relations [6.35] and [6.36], the general expression of the isobar becomes as follows, if we ignore the variations in isosteric heat with changing temperature:

$$\frac{q_{isost}}{\mathrm{R}T^2} = -\left(\frac{\partial \mu_{\langle(G)\rangle}}{\partial n_a}\right)_{P,T,n^{(A)}_{\langle(S)\rangle}} \mathrm{d}n_a = 0 \qquad [6.41]$$

The choice of a solution model will, as before, enable us to calculate the differential of the chemical potential in relation to the composition.

6.3.2. *Hill and Everett's model*

The adsorbed phase is assimilated to a pure condensed phase of the species forming the gas, which is represented as $\langle(G)\rangle$. This phase could, for instance, be characterized by a function of state. In order to calculate the Gibbs variance, we consider that we have a single component G in two different phases – one gaseous, {G}, and the other condensed, $\langle(G)\rangle$ – and that there are three external intensive factors: temperature, pressure and the interaction due to the solid surface. We then see a situation very similar to the liquid film deposited on a solid, as seen previously (see section 5.4.4).

6.3.2.1. *General equilibrium equation*

If we let A_M denote the area occupied by a mole on the surface, and $s^0_{\langle(G)\rangle}$ and $v^0_{\langle(G)\rangle}$ denote the molar entropy and molar volume of the pure adsorbed phase, the differential of the molar Gibbs energy of the adsorbed layer is:

$$\mathrm{d}g^0_{\langle(G)\rangle} = -s^0_{\langle(G)\rangle}\,\mathrm{d}T + v^0_{\langle(G)\rangle}\,\mathrm{d}P + A_M\,\mathrm{d}\Phi \qquad [6.42]$$

We let Φ denote the value defined by the expression:

$$\Phi = \sigma_{\mathrm{SG}} - \sigma_{\mathrm{AG}} \tag{6.43}$$

σ_{SG} and σ_{AG} are, respectively, the interfacial tensions between the initial solid phase and the gaseous phase and between the adsorbed phase and the gaseous phase.

We can see that in expression [6.42], for the surface phase, in relation to the area A, Φ plays the same role as pressure in a three-dimensional phase in relation to the volume. Thus, Φ is called the *expansion pressure*.

Let Γ denote the inverse of the molar area. Γ tells us the number of moles adsorbed per unit area, defined by relation [2.1]. This number is expressed, in our particular case, by:

$$\Gamma = \frac{1}{A_M} \tag{6.44}$$

Relation [6.42] is then written:

$$\mathrm{d}\, g^0_{\langle(\mathrm{G})\rangle} = -s^0_{\langle(\mathrm{G})\rangle}\, \mathrm{d}T + v^0_{\langle(\mathrm{G})\rangle}\, \mathrm{d}P + \frac{1}{\Gamma}\mathrm{d}\Phi \tag{6.45}$$

With regard to the compound G in the gaseous phase, its molar Gibbs energy is:

$$\mathrm{d}\, g^0_{\{\mathrm{G}\}} = -s^0_{\{\mathrm{G}\}}\, \mathrm{d}T + v^0_{\{\mathrm{G}\}}\, \mathrm{d}P \tag{6.46}$$

Equality between the two differentials [6.45] and [6.46] gives us the general equilibrium equation:

$$\left(s^0_{\{\mathrm{G}\}} - s^0_{\langle(\mathrm{G})\rangle}\right)\mathrm{d}T + \left(v^0_{\{\mathrm{G}\}} - v^0_{\langle(\mathrm{G})\rangle}\right)\mathrm{d}P + \frac{1}{\Gamma}\mathrm{d}\Phi = 0 \tag{6.47}$$

From this equation, we can deduce the different functions with a single variable – particularly the isotherm and the isostere.

6.3.2.2. *Equation of the isotherm*

At a constant temperature, the general equation [6.47] is reduced to:

$$\left(v_{\{G\}}^{0}-v_{\langle\langle G\rangle\rangle}^{0}\right)\mathrm{d}P+\frac{1}{\Gamma}\mathrm{d}\Phi=0 \qquad [6.48]$$

In light of the usual approximations about the volumes:

$$v_{\{G\}}^{0}=\frac{\mathrm{R}T}{P}>>v_{\langle\langle G\rangle\rangle}^{0} \qquad [6.49]$$

The equation of the isotherm is written:

$$\mathrm{d}\Phi=-\mathrm{R}T\Gamma\,\mathrm{d}\ln P \qquad [6.50]$$

We can also write this in its integral form:

$$\Phi=-\mathrm{R}T\int_{0}^{P}\Gamma\,\mathrm{d}\ln P \qquad [6.51]$$

This relation is *Gibbs' relation for adsorption*. In particular, it is the case of relation [2.41] applied to the adsorption of a pure gas by a pure solid.

6.3.2.3. *Equation of the isostere: equilibrium heat of adsorption*

We now work with the spreading constant ($\mathrm{d}\Phi = 0$). The general equation [6.47] becomes:

$$\left(s_{\{G\}}^{0}-s_{\langle\langle G\rangle\rangle}^{0}\right)\mathrm{d}T+\left(v_{\{G\}}^{0}-v_{\langle\langle G\rangle\rangle}^{0}\right)\mathrm{d}P=0 \qquad [6.52]$$

The equation of the isostere is then written:

$$\left(\frac{\mathrm{d}P}{dT}\right)_{\Phi}=\frac{s_{\{G\}}^{0}-s_{\langle\langle G\rangle\rangle}^{0}}{v_{\{G\}}^{0}-v_{\langle\langle G\rangle\rangle}^{0}}=\frac{s_{\{G\}}^{0}-s_{\langle\langle G\rangle\rangle}^{0}}{\mathrm{R}T}P \qquad [6.53]$$

Because of the equality of the molar Gibbs energies at equilibrium, we can introduce the enthalpy terms into that expression, and write the isostere in the form:

$$\left(\frac{\mathrm{d}\ln P}{\mathrm{d}T}\right)_{\Phi}=\frac{s^{0}_{\{G\}}-s^{0}_{\langle(G)\rangle}}{RT}=\frac{h^{0}_{\{G\}}-h^{0}_{\langle(G)\rangle}}{RT^{2}}=\frac{q_{\Phi}}{RT^{2}} \qquad [6.54]$$

where q_{Φ} is the equilibrium heat of adsorption.

We can link the molar enthalpy of the adsorbed phase to the molar enthalpy of the liquid component G by writing that the difference between the two phases is due only to the surface effect, and that this surface effect has no impact on the entropy, so:

$$h^{0}_{\langle(G)\rangle}=h^{0}_{(G)}+\Phi A_{M} \qquad [6.55]$$

Hence, as the entropies of the liquid state and the adsorbed state are identical, we have:

$$s^{0}_{\langle(G)\rangle}=s^{0}_{(G)} \qquad [6.56]$$

The two states are very similar, and the adsorbed state is, in fact, assimilated to a thin liquid film, as we saw in section 5.4.4.

6.3.3. *Adsorption heats*

We have already defined two heat values linked to adsorption: the isosteric heat, based on Hill's model, given by:

$$q_{\text{isost}}=RT^{2}\left(\frac{\mathrm{d}P}{dT}\right)_{n_a} \qquad [6.57]$$

and the adsorption heat based on Hill and Everett's model, given by the relation:

$$q_{\Phi}=RT^{2}\left(\frac{\mathrm{d}P}{dT}\right)_{\Delta\sigma} \qquad [6.58]$$

The first is easy to determine experimentally by finding the slope of the isostere when the adsorbed quantity is constant. In order to calculate the second, we shall now establish a relation between these two adsorption heats. These two heats are defined, on the basis of the entropies, by relations [6.35] and [6.54], respectively. By adding these two definitions together, term by term, and adding and subtracting the quantity $\overline{S_{\langle\langle G\rangle\rangle}}$ from the relation thus obtained, we find:

$$q_{\Phi} = \left(s^{0}_{\{G\}} - s^{0}_{(G)} - \overline{S_{\langle\langle G\rangle\rangle}} + \overline{S_{\langle\langle G\rangle\rangle}}\right)T = q_{\text{isost}} + \left(\overline{S_{\langle\langle G\rangle\rangle}} - s^{0}_{(G)}\right)T \qquad [6.59]$$

However, the differential of the chemical potential at constant pressure, area and adsorbed quantity of the adsorbed species is independent of the model. Thus, we can write the equality between that chemical potential given by relation [6.24] for Hill's model and that given by relation [6.45] for Hill and Everett's model:

$$\left(\mathrm{d}\mu_{\{G\}}\right)_{P,A,n_a} = -S_{\langle\langle G\rangle\rangle}\,\mathrm{d}T = -s^{0}_{(G)}\,\mathrm{d}T + \frac{1}{\Gamma}\mathrm{d}\Phi \qquad [6.60]$$

From this, we deduce:

$$\left(\overline{S_{\langle\langle G\rangle\rangle}} - s^{0}_{(G)}\right) = -\frac{1}{\Gamma}\left(\frac{\partial \Phi}{\partial T}\right)_{\Gamma} \qquad [6.61]$$

and by substituting back into relation [6.59], it follows that we have:

$$q_{\Phi} = q_{\text{isost}} - \frac{T}{\Gamma}\left(\frac{\partial \Phi}{\partial T}\right)_{\Gamma} \qquad [6.62]$$

Generally speaking, there are as many heats linked to adsorption as there are ways to achieve it, because the heat exchanged in the course of a transformation is not a function of state. Let Q denote such a heat defined in specific conditions. We speak of the corresponding *differential heat* to refer to the differential of Q in relation to the adsorbed quantity:

$$q = \frac{\mathrm{d}Q}{\mathrm{d}n_a} \qquad [6.63]$$

The two quantities q_{isost} and q_a are two examples of differential heats linked to adsorption.

Thus, for each differential heat, there will be a corresponding integral heat, defined by:

$$Q = \int_0^{n_a} q \, \mathrm{d} n_a \qquad [6.64]$$

A commonplace mode of adsorption is adsorption where the volume of the gas, the volume of the quantity adsorbed and the area of the solid are kept constant. Then, the differential of the heat is given by the difference between the Helmholtz energies:

$$(\mathrm{d}Q)_{V_{\{G\}},\, V_a, A} = \mathrm{d}U_{\{G\}} - \mathrm{d}U_a \qquad [6.65]$$

Consider the case of Hill's model. For the Helmholtz energies, we have:

$$u^0_{\{G\}} = \frac{\mathrm{d}U_{\{G\}}}{\mathrm{d}n_{\{G\}}} \qquad [6.66a]$$

and:

$$\overline{U_{\langle\langle G \rangle\rangle}} = \frac{\mathrm{d}U_a}{\mathrm{d}n_a} \qquad [6.66b]$$

The corresponding differential heat, then, will be:

$$\frac{(\mathrm{d}Q)_{V^0_{\{G\}}, V_a, A}}{\mathrm{d}n_a} = u^0_{\{G\}} - \overline{U_{\langle\langle G \rangle\rangle}} = q_d \qquad [6.67]$$

q_d is called the differential heat of adsorption: it is the heat that would be measured when carrying out adsorption in an enclosed isothermal calorimeter, such as the Calvet microcalorimeter.

The differential heat of adsorption can be linked to the isosteric heat by finding the difference of these two heats, and by using the entropic definition [6.35] for the latter, we find:

$$q_d - q_{\text{isost}} = u^0_{\{G\}} - Ts^0_{\{G\}} - \left(\overline{U_{\langle\langle G\rangle\rangle}} - T\overline{S_{\langle\langle G\rangle\rangle}}\right) \qquad [6.68]$$

We know that generally, given the definition of the Gibbs energy, we can write:

$$U - TS = G - PV \qquad [6.69]$$

When applied to relation [6.68], this gives us:

$$q_d - q_{\text{isost}} = g^0_{\{G\}} - Pv^0_{\{G\}} - \left(\mu_{\langle\langle G\rangle\rangle} - P\overline{V_{\langle\langle G\rangle\rangle}}\right) \qquad [6.70]$$

In addition, at equilibrium, we have equality of the chemical potentials in the gas phase and the adsorbed phase. Hence, in Hill's model:

$$g^0_{\{G\}} = \mu_{\langle\langle G\rangle\rangle} \qquad [6.71]$$

By feeding this back into expression [6.70], we obtain:

$$q_d - q_{\text{isost}} = -P\left(v^0_{\{G\}} - \overline{V_{\langle\langle G\rangle\rangle}}\right) \qquad [6.72]$$

With the usual approximation about the relative values of the molar volumes of the adsorbed phase and the gaseous phase, the latter being considered to be a perfect gas, we find:

$$q_d - q_{\text{isost}} = -\text{R}T \qquad [6.73]$$

Because the two values q_d and q_{isost} can be measured experimentally, this relation is easy to verify.

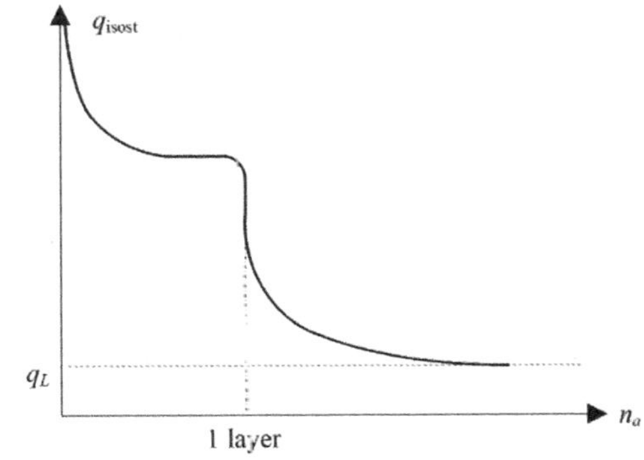

Figure 6.4. *Example of the variation in the isosteric heat with the adsorbed quantity*

Many experimental results can be explained by the presence of van der Waals forces and by relation [6.7].

In general, the adsorption heats are not constant: they depend on the quantity of material adsorbed, although graphite is an exception to this rule. This variation with the adsorbed quantity is proof that there is heterogeneity over the course of the adsorption. Figure 6.4 shows an example of such a heterogeneity.

Note, though, that beyond a certain adsorbed quantity, the heat of adsorption is practically equal to the enthalpy of liquefaction of the gas $\Delta_{\mathrm{v}\rightarrow\mathrm{L}}H$.

6.4. Monolayer adsorption

Adsorption with the formation of a single layer of gas adsorbed stems from the Type I isotherm in Brunauer's classification (see Figure 6.1). The adsorbed quantities are relatively slight: to saturate a surface, it takes around 10^{15} molecules per cm^2, which equates to around 10^{-4} moles per m^2. Saturation is reached for low values of the ratio P/P^0, far from the liquid state. In this case, the concept of the adsorbed quantity is often replaced by that of the fraction of coverage, θ, which is defined by the ratio of the surface of the solid covered by the gas to the total surface of the solid.

6.4.1. *Energy distribution of adsorbed molecules*

The two expressions, [6.39] and [6.54], of the isotherms given by Hill's and Hill and Everett's models cannot be immediately used to plot the isothermal curves without further information about the adsorbed state.

In section 6.1, we calculated mean energies over the whole surface for the adsorption of a gas. We can see that the periodicity of the arrangements of the atoms on the surface of a solid results, in fact, in a periodic energy of adsorption, exhibiting points of minimum energy, in a position of stable equilibrium, and points of maximum energy in a position of unstable equilibrium. The difference between the maximum and minimum energies, therefore, needs to be compared to the heat energy k_BT (Figure 6.5). Two limiting cases are found:

– if the difference between the energies of two neighboring positions of stable equilibrium is greater than the energy of thermal agitation (Figure 6.5(a)), then that phenomenon prevents the molecule from crossing the barriers between two positions. We then say that we have *localized adsorption*;

– if, on the other hand, the difference between the energies of two neighboring positions of stable equilibrium is smaller than the energy of thermal agitation (Figure 6.5(b)), it enables the molecule to cross the barriers between two positions, in which case we say we have *mobile adsorption*.

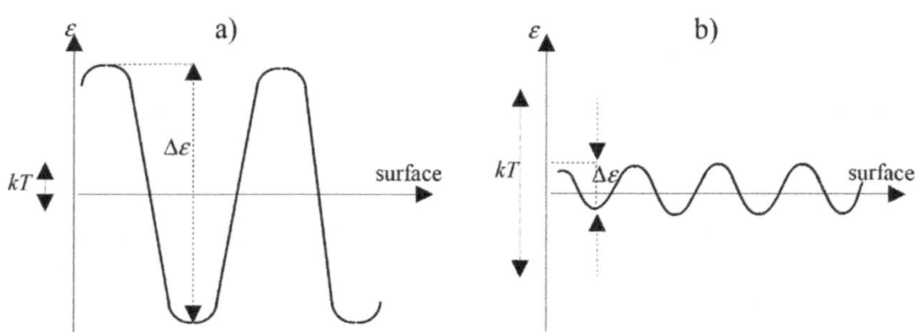

Figure 6.5. *Energy states of adsorbed molecules: a) localized; b) mobile*

In addition, we have seen that the heats of adsorption varied with the adsorbed quantity, even during the course of the adsorption of a single layer (Figure 6.4). This means that when a certain number of molecules have been adsorbed, the adsorption of a new molecule does not involve the same variation in energy as the adsorption of the previous ones. This can be interpreted in two different ways:

– either the surface of the solid is not energetically homogeneous. Our calculation of this energy in section 6.2 was a mean value, but this value may fluctuate from one point on the surface to another;

– or there is an interaction between the molecules already adsorbed, which is expressed by the fact that the potential energy of a given site depends on the state of occupation of the neighboring sites.

The calculation of the isotherm in the two cases is different, but the result of the experiment does not enable us to distinguish between these two cases, because on the adsorption of a new molecule, there is no way of telling whether the energy variation seen already existed from the start on the bare surface, or whether it is caused by the presence of the molecules adsorbed previously. We say, in both cases, that we have *adsorption with interactions*.

Depending on the different properties – layers localized or otherwise, with or without interactions between the molecules – we shall examine the different models used to describe monolayer adsorption.

6.4.2. *Isotherms of adsorption in mobile monolayers with no interaction*

The adsorbed molecules, arranged in a monolayer, are free to move around on the surface, and there is no interaction between them.

6.4.2.1. *Hill and Everett's model*

In this model, the adsorbed phase is a pure, two-dimensional phase, for which we can calculate the molecular partition function with only two degrees of freedom. It is of the form:

$$z_a = A\left(\frac{2\pi m \mathrm{k_B} T}{\mathrm{h}^2}\right) z_{\mathrm{int}} \qquad [6.74]$$

In this relation, z_{int} is the contribution, to the molecular partition function, of the energy terms internal to the molecule. The term in parentheses is the contribution of a two-dimensional motion of translation.

If we look at the limiting statistical case for N_{ads} indiscernible molecules for a single gas, the molecular partition function is:

$$Z_a = \frac{z_a}{N_{ads}!} \qquad [6.75]$$

The Helmholtz energy of the adsorbed molecules is given by:

$$F_a = -\mathrm{N_a k_B} T \ln Z_a \qquad [6.76]$$

By applying Stirling's approximation and combining relations [6.74], [6.75] and [6.76], the Helmholtz energy of the adsorbed molecules is expressed by:

$$F_a = -\mathrm{N_a k}_B T \ln\left\{\frac{N_{ads}\mathrm{h}^2}{2\pi m \mathrm{k_B} T z_{\mathrm{int}} A}\right\} - \mathrm{N_a k_B} T \qquad [6.77]$$

From this, we deduce, by analogy with a pressure, the pressure of expansion by differentiating the Helmholtz energy in relation to the surface:

$$\Phi = -\frac{\partial F_a}{\partial A} = \frac{\mathrm{N_a k_B} T}{A} \qquad [6.78a]$$

or:

$$\Phi = \Gamma \mathrm{R} T \qquad [6.78b]$$

Equation [6.78b] is an equation of state for the adsorbed phase.

NOTE.– It should be noted that this equation of state is for a two-dimensional phase, similar to that of the perfect gas for a bulk phase, and is written as follows for one mole of adsorbed molecules:

$$\Phi A_M = \mathrm{R} T \qquad [6.79]$$

By looking again at the differential form [6.50] of Gibbs' equation and differentiating relation [6.78b], we obtain:

$$\mathrm{R} T \,\mathrm{d}\Gamma = \mathrm{R} T \Gamma \,\mathrm{d} \ln P \qquad [6.80]$$

which, after integration, gives us:

$$\Gamma = kP \tag{6.81}$$

This isotherm is called *Henry's isotherm* because it is similar to Henry's law of solubility of perfect gases in a liquid.

6.4.2.2. *Hill's model*

Looking at relation [6.39] again and assuming the solution is perfect – i.e. for the chemical potential, we have the expression:

$$\mu_{<<G>>} = \mu^0_{<<G>>} + RT \ln \Gamma \tag{6.82}$$

By differentiating relation [6.82] and feeding back into expression [6.39], we find:

$$\mathrm{d} \ln P = \mathrm{d} \ln \Gamma \tag{6.83}$$

which again gives us Henry's isotherm [6.81].

Thus, Hill and Everett's and Hill's models yield the same expression for the isotherm of adsorption.

Instead of applying relation [6.82] for the perfect gas, we could have repeated the whole reasoning process by statistical thermodynamics for a perfect solution, transposed to a two-dimensional phase, and of course, we would have again obtained Henry's expression.

6.4.3. *Isotherms of adsorption in mobile monolayers with interactions*

Let us again consider Hill and Everett's model. There are two methods available to us to take account of the interactions between molecules in the adsorbed phase:

– either we adopt a statistical description, formulating the model developed in the case of gases and thus transposing the second coefficient of the virial to a two-dimensional phase;

– or we choose to transpose an equation of state for real gases and apply it to the adsorbed phase.

Let us choose the second method and define the surface phase by an equation similar to van der Waals' equation for gases:

$$\Phi\left(A_M - b\right) = \mathrm{R}T \qquad [6.84]$$

Here, b is the co-surface – i.e. the surface area covered when it is completely saturated. Obviously, for the concentrations, we have:

$$\Gamma = \frac{1}{A_M} \qquad [6.85a]$$

and:

$$\Gamma_\infty = \frac{1}{b} \qquad [6.85b]$$

By feeding back into expression [6.84], we find:

$$\Phi = \frac{\mathrm{R}T\Gamma}{1 - \dfrac{\Gamma}{\Gamma_\infty}} \qquad [6.86]$$

By differentiating, we obtain:

$$\mathrm{d}\Phi = \mathrm{R}T \frac{d\Gamma}{\left(1 - \dfrac{\Gamma}{\Gamma_\infty}\right)^2} \qquad [6.87]$$

By identifying with Gibbs' equation [6.50], we are led to:

$$\Gamma d \ln P = \frac{d\Gamma}{\left(1 - \dfrac{\Gamma}{\Gamma_\infty}\right)^2} \qquad [6.88]$$

Knowing that the coverage fraction is given by:

$$\frac{\Gamma}{\Gamma_\infty} = \theta \qquad [6.89]$$

By integrating [6.88] in light of [6.89], we obtain:

$$P = k\frac{\theta}{1-\theta}\exp\left(\frac{\theta}{1-\theta}\right) \qquad [6.90]$$

This is the *Volmer isotherm*.

If we replace van der Waals' equation of state [6.84] with Berthelot's, written in the form:

$$\left(\Phi + \frac{a}{A_M^2}\right)(A_M - b) = \mathrm{R}T \qquad [6.91]$$

the same calculation gives us:

$$P = k\frac{\theta}{1-\theta}\exp\left(\frac{\theta}{1-\theta}\right) - \frac{2a\Gamma_\infty\theta}{\mathrm{R}T} \qquad [6.92]$$

This Type I isotherm was found by Hill using statistical thermodynamics. It is true that for all equations of state of gases that we are able to find by statistical thermodynamics, we can replace the macroscopic reasoning on the basis of the equation of state with the corresponding microscopic demonstration on the basis of statistical thermodynamics. Not every approach, though, constitutes a new model, as we find all too often in the existing body of literature.

6.4.4. *Isotherms of adsorption in localized monolayers without interaction*

The adsorbed molecule is attached at a given point on the surface, and we consider it to be fixed to an adsorption site. The surface of the solid can then be viewed as a lattice of free or occupied sites. The model is said to be "without interaction" when the interactions between free sites and occupied sites or between different occupied sites are identical to the interactions between different free sites.

We shall work on the basis of a slight modification of Hill's model, considering that the surface is a solution of free sites $\langle\langle \mathrm{s} \rangle\rangle$ and occupied sites $\langle\langle \mathrm{SG} \rangle\rangle$. In the model without interaction, we find ourselves precisely in the conditions of a perfect solution.

The adsorption reaction is written:

$$\langle\langle \mathrm{S} \rangle\rangle + \{G\} = \langle\langle SG \rangle\rangle \qquad [6R.1]$$

Considering the solution to be perfect, the law of mass action gives us:

$$K_a = \frac{x_{\langle\langle \mathrm{SG} \rangle\rangle}}{x_{\langle\langle \mathrm{S} \rangle\rangle} P} \qquad [6.93]$$

It is easy to link the molar fractions to the degree of coverage of the surface. We have:

$$x_{\langle\langle \mathrm{SG} \rangle\rangle} = \frac{s_0 - s}{s_0} = \theta \qquad [6.94a]$$

$$x_{\langle\langle \mathrm{S} \rangle\rangle} = \frac{s}{s_0} = 1 - \theta \qquad [6.94b]$$

By feeding these expressions back into equation [6.93], we find:

$$\theta = \frac{K_a P}{1 + K_a P} \qquad [6.95]$$

The coverage fraction is a homographic function of the gas pressure. It is the equation of the *Langmuir isotherm* for Type I isotherms.

6.4.5. *Isotherms of adsorption in localized monolayers with interactions*

We can model this scenario either by introducing a macroscopic model of the solution which allows us to give the expressions of the activities of the components $\langle\langle \mathrm{S} \rangle\rangle$ and $\langle\langle \mathrm{SG} \rangle\rangle$, or by going back to the source and imagining a statistical model, as we did to find certain macroscopic models of solutions.

We shall use the first method, employing the model of the strictly-regular solution.

Instead of relation [6.93], the law of mass action is expressed as a function of the activities in the form:

$$K_a P = \frac{x_{\langle\langle SG\rangle\rangle}\gamma_{\langle\langle SG\rangle\rangle}}{x_{\langle\langle S\rangle\rangle}\gamma_{\langle\langle S\rangle\rangle}} \qquad [6.96]$$

If the surface solution is strictly regular, the activity coefficients are expressed in the form:

$$\ln \gamma_{\langle\langle S\rangle\rangle} = \frac{B}{T}\theta^2 \qquad [6.97a]$$

$$\ln \gamma_{\langle\langle SG\rangle\rangle} = \frac{B}{T}(1-\theta)^2 \qquad [6.97b]$$

By feeding those expressions back into relation [6.96], we find:

$$K_a P = \frac{\theta}{1-\theta}\exp\frac{B}{T}\exp-\frac{2B\theta}{T} \qquad [6.98]$$

If we switch to a microscopic model using statistical thermodynamics, we obviously obtain the same result as for strictly-regular solutions (see Chapter 3, of [SOU 15b]), so for the coefficient B, we have the expression:

$$B = \frac{\mathrm{N}_a w_{S,GS}}{k_B} \qquad [6.99]$$

The interaction term $w_{S,GS}$ is, itself, defined by applying the expression:

$$w_{\mathrm{S,GS}} = z\left(\varepsilon_{S,GS} - \frac{\varepsilon_{S,S} + \varepsilon_{GS,GS}}{2}\right) \qquad [6.100]$$

The isotherm [6.98] is known as the *Fowler isotherm*.

NOTE.– The fact that we are dealing with a two-dimensional solution, rather than a three-dimensional one, is reflected in the expression of the constant K_a, which contains the partition function due to two translational terms instead of three.

The presence of interactions is, as stated earlier, either the consequence of a genuine interaction between the adsorbed molecules or the result of a non-uniform surface – i.e. one which is heterogeneous in terms of energy. In order to continue using the relation of the Langmuir isotherm, Graham proposed to keep this expression but replace the equilibrium constant K_a by a variable K_f. Thus, K_f is called the *equilibrium function*. The variations of that function will depend on the degree of coverage in accordance with a curve, which depends on the type of adsorption.

Figure 6.6 shows a few examples of variations in K_f.

For a layer without interaction, $K_f = K_a$ is a constant (this is the Langmuir isotherm): curve **a** in Figure 6.6.

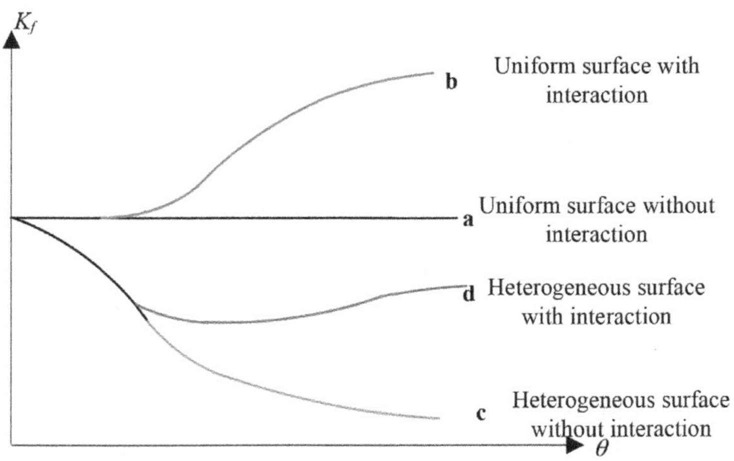

Figure 6.6. *Equilibrium function in the different cases of interactions*

For a non-ideal system, we may have:

– interactions between molecules adsorbed to a uniform surface (curve b in Figure 6.6), as is the case, for instance, with the Fowler isotherm [6.98], for which we obtain:

$$K_f = K_a \exp - \frac{B}{T} \exp \frac{2B\theta}{T} \qquad [6.101]$$

– adsorption on a heterogeneous surface *without interaction* (curve c in Figure 6.6);

– adsorption on a heterogeneous surface *with interactions* between the adsorbed molecules (curve d in Figure 6.6).

In conclusion, we can see that the number of isotherm equations likely to represent the monolayer is limitless by the introduction of a varied equation of state for mobile layers or of different solution models for immobile adsorption layers.

6.5. Multilayer adsorption

We now turn our attention to models capable of accounting for the shapes of Type II and Type III isotherms (see Figure 6.1). In this case, the adsorbed quantity is much greater than in the case of a monolayer, and a certain saturation occurs when the ratios P/P^0 are near to 1.

This multilayer adsorption is such that we can consider the first layers to be successive monolayers, in which the forces exerted by the solid are crucially important, but as we move further away from the solid, its influence becomes steadily less and becomes slight in comparison to the influence of the lower layers, until the liquid phase is reached, at which point only the adsorbed molecules exert a force on the last molecules to attach. We can then see why, in order to model such a continuous variation of the properties, two limiting cases were examined. The Brunauer, Emmet and Taylor model favors the multi-monolayer aspect, and is valid for pressures far from the value of condensation into liquid, whilst the Frankel, Halsey and Hill, and Polanyi models favor the neighboring properties of the liquid for the adsorbed layer and generally perform better as we approach the pressure of condensation into liquid.

6.5.1. *The Brunauer, Emmet and Taylor (B.E.T.) isotherm*

The adsorption is localized, based on Langmuir's model, but a given site may be covered by multiple layers, and only the last adsorbed layer, furthest away from the solid, is at equilibrium with the gases.

A given site may be covered by 0, 1, 2 … localized layers of gas without interaction. Let A_0, A_1, A_2, …, A_i represent the areas covered respectively by 0, 1, 2, …, i layers of gas (Figure 6.7).

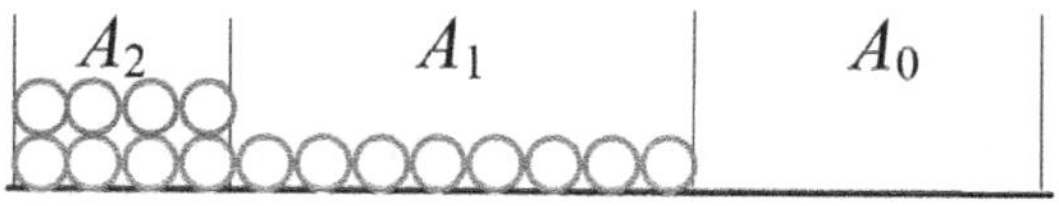

Figure 6.7. *Filling by layers in the B.E.T. model*

If A is the total area of the solid, the equilibrium between the bare surface and the surface covered by a layer means we can write:

$$\frac{A_1/A}{A_0/A} = \frac{A_1}{A_0} = K_1 P \qquad [6.102]$$

The equilibrium between the surface covered by a layer and the surface covered by two layers is expressed by:

$$\frac{A_2}{A_1} = K_2 P \qquad [6.103]$$

and so on. The equilibrium between the layers i and i–1 gives us:

$$\frac{A_i}{A_{i-1}} = K_i P \qquad [6.104]$$

Let us state that the sum of the surfaces is equal to the total surface:

$$\sum_{i=0}^{\infty} A_i = A \qquad [6.105]$$

In addition, if we let v_0 denote the volume of gas per unit surface which would just cover the whole surface of a monolayer, the total volume would be:

$$V = 0 + 1.v_0 A_1 + 2.v_0 A_2 ... + i.v_0 A_i + ... = v_0 \sum_{i=0}^{\infty} iA_i \qquad [6.106]$$

Let v_{mono} denote the product:

$$v_{\text{mono}} = v_0 S_t \qquad [6.107]$$

v_{mono} thus denotes the volume of gas which would be just necessary to cover the whole surface with a monolayer.

The total volume found can be written:

$$V = \frac{v_{\text{mono}}}{A} \sum_{i=0}^{\infty} iA_i = v_{\text{mono}} \frac{\sum_{i=0}^{\infty} iA_i}{\sum_{i=0}^{\infty} A_i} \qquad [6.108]$$

To find a simple solution, we shall suppose that from the point of fixation of the second layer, with adsorption always taking place on a previously-adsorbed layer, the equilibrium constant remains the same – i.e. $K_2 = K_3 = \ldots = K_i = \ldots = K_a$.

To simplify the formulae, let us posit that:

$$x = \frac{P}{K_a} \qquad [6.109a]$$

and by analogy:

$$cx = \frac{P}{K_1} \qquad [6.109b]$$

with the relation:

$$c = \frac{K_a}{K_1} \qquad [6.110]$$

The system of equations for the equilibria then becomes:

$$\begin{cases} A_1 = cxA_0 \\ A_2 = cx^2 A_0 \\ \dots\dots\dots\dots \\ A_i = cx^i A_0 \\ \dots\dots\dots\dots \end{cases}$$

By finding the sum, member by member:

$$A = A_0\left[1 + c\left(x + x^2 + \ldots + x^i + \ldots\right)\right] \qquad [6.111]$$

Supposing, for the moment, that $0 < x < 1$ (this inequality is explained and justified a little later on), relation [6.111] becomes:

$$A = A_0\left(1 + \frac{cx}{1-x}\right) = A_0\left[1 + \frac{(c-1)x}{1-x}\right] \qquad [6.112]$$

In the same way, we calculate:

$$\sum_{i=0}^{\infty} iA_i = cxA_0\left(1 + +2x + 3x^2 + \ldots + ix^{i-1} + \ldots\right) = \frac{cxA_0}{(1-x)^2} \qquad [6.113]$$

By substituting back into equation [6.108], the fixed volume would be:

$$V = \frac{cxv_{\text{mono}}}{(1-x)\left[1 + (c-1)x\right]} \qquad [6.114]$$

Let us look for the physical meaning of x. If $x = 1$, the volume tends toward infinity, meaning that the gas liquefies, and thus the pressure is: $P = P^0$. Hence:

$$\frac{P^0}{K_1} = 1 \quad [6.115a]$$

Thus:

$$x = \frac{P}{P^0} < 1 \quad [6.115b]$$

The expression of the isotherm, if we feed back this final equation into relation [6.114], is written:

$$V = v_{\text{mono}} \frac{cP}{\left(P^0 - P\right)\left[1 + (c-1)\frac{P}{P^0}\right]} \quad [6.116]$$

This is the equation of the *B.E.T. isotherm*.

This equation can also be written:

$$\frac{P}{V\left(P^0 - P\right)} = \frac{1}{cv_{\text{mono}}} + \frac{(c-1)P/P^0}{cv_{\text{mono}}} \quad [6.117]$$

Experimentally, if on the basis of the measurement of the volume adsorbed at each pressure, we plot the left-hand side of equation [6.117] as a function of the ratio P/P^0, we obtain a straight line with the slope $1/cv_{\text{mono}}$ and with ordinate at the origin $(c-1)/cv_{\text{mono}}$, which is why it is possible to calculate values for v_{mono} and c.

If we establish the B.E.T. relation using statistical thermodynamics, the constant c appears to be the ratio:

$$c = \exp\frac{|\Delta_1 H| - |\Delta_{\text{v}\to\text{L}} H|}{\text{R}T} \quad [6.118]$$

In this expression, $\Delta_1 H$ is the enthalpy of adsorption of the first layer of gas, and $\Delta_{\mathrm{v}\to\mathrm{L}} H$ is the enthalpy of liquefaction of the gas.

NOTE.– The enthalpy of adsorption is less strongly negative than the enthalpy of liquefaction of the gas, so we deduce that the value c must be positive. Consequently, if in a given experiment we are led to adopt a negative value of c, then we shall be led to reject the B.E.T. model. The coincidence of the isotherm with relation [6.116] is fortuitous.

Note that the expression of the isotherm is established for an infinite number of layers, which accounts for the name *infinite form of the B.E.T. equation* given to relation [6.116]. The statistical calculation carried out for a finite number n of layers leads to:

$$V = v_{\text{mono}} \frac{cx}{1-x} \frac{1-(n+1)x^n + nx^{n+1}}{1+(c-1)x - cx^{n+1}} \qquad [6.119]$$

This is the *B.E.T. isotherm with n layers*.

NOTE.– The model accepts the coexistence of a high number of layers and of a non-covered surface. This, in fact, leads to negative surface tensions, which is the reason why the model only conforms to real-world experience when the pressure ratio P/P^0 is less than 0.35.

6.5.2. *Frenkel, Halsey and Hill's liquid layer model*

This model is founded on Hill and Everett's hypothesis. The adsorbed layer is compared to a liquid. The two phases are subject to van der Waals forces deriving from a potential given by relation [6.7], and the hypothesis is adopted that the difference between the adsorbed layer and the liquid lies only in the difference between their van der Waals potentials. The reference liquid has a uniform thickness h and contains Γ moles per unit surface. It has a density of $N_{(\mathrm{L})}$ moles per unit volume such that:

$$N_{(\mathrm{L})} = \frac{\Gamma}{h} \qquad [6.120]$$

This difference in van der Waals potentials is, in view of relation [6.7], and if we ignore the effects of repulsion:

$$\Delta\varepsilon^{(G)} = \frac{\pi N \varepsilon_0 r_0^6}{3z^3} - \frac{\pi N_{(L)} \varepsilon_0 r_{0(L)}^6}{3z^3} \tag{6.121}$$

In light of the pressures of equilibrium with the gas, the chemical potential difference of the fixed species between the adsorbed phase and the liquid is:

$$\mu - \mu_{(L)} = RT \ln \frac{P}{P^0} \tag{6.122}$$

We consider that this difference is due only to the differences in Helmholtz energy – i.e. to the difference between the van der Waals potentials. By comparing relations [6.121] and [6.122] and replacing h with its value, drawn from expression [6.120], we obtain:

$$\ln \frac{P}{P^0} = -\frac{a}{\Gamma^3} \tag{6.123}$$

We know that a is constant at a given temperature, which is expressed thus:

$$a = \frac{\pi N_{(L)}^4 \varepsilon_0 r_{0(L)}^6 - \pi N N_{(L)}^3 \varepsilon_0 r_0^6}{3k_B T} \tag{6.124}$$

Other authors have proposed adopting a more general expression than relation [6.123], in the form:

$$\ln \frac{P}{P^0} = -\frac{a}{\Gamma^n} \tag{6.125}$$

where $2 < n < 3$.

By another method, based on Hill and Everett's thermodynamic model, Harkins and Jura found $n = 2$.

6.5.3. *Polanyi's potential model*

The model put forward by Polanyi is very old (1914) but remains useful even today, as it is the only model which is capable of making predictions.

Relying once again on Hill and Everett's hypothesis, we perform the same comparison as in the previous model, attributing the difference between the liquid and the adsorbed phase simply to the difference in interaction potential between the solid and the molecule, so we can write:

$$\mu_{(L)} - \mu = RT \ln \frac{P^0}{P} = \Delta\varepsilon \qquad [6.126]$$

Figure 6.8(a) shows the plot of the equipotential curves ($\Delta\varepsilon$ constant) in the vicinity of the surface of a solid.

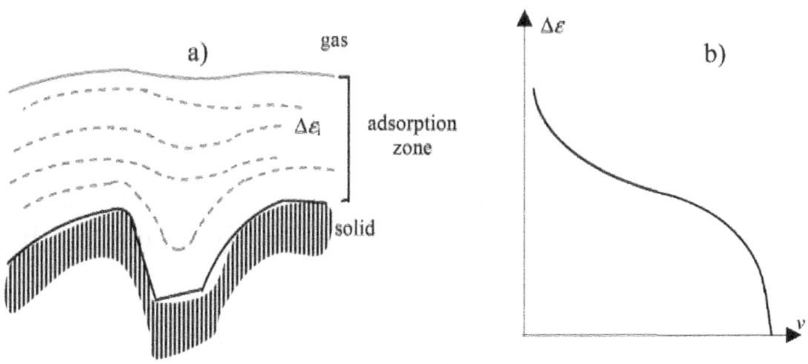

Figure 6.8. *Polanyi's potential model: a) equipotentials; b) characteristic curve*

This time, we shall not state an expression for $\Delta\varepsilon$, instead choosing a different means of comparison.

Let v_1, v_2, ..., v_{max} denote the volumes contained between the solid and the equipotentials $\Delta\varepsilon_1$, $\Delta\varepsilon_2$, ..., 0. v_{max} is the total volume of the adsorption zone. As v increases from 0 to v_{max}, $\Delta\varepsilon$ decreases from its maximum value to 0.

The process of construction of the adsorbed layer can then be represented by a curve known as the characteristic adsorption curve $\Delta\varepsilon = f(v)$. That curve can be plotted on the basis of an isotherm found experimentally, by way of a hypothesis about the nature of the adsorbed phase which, in our case, is supposed to be very similar to the liquid. In this case, the volume v_i corresponding to a potential $\Delta\varepsilon_i$ will be given by:

$$v_i = \frac{m_i}{\rho_{(L)}} \tag{6.127}$$

where m_i is the adsorbed mass and $\rho_{(L)}$ is the density of the liquid at temperature T of the experimental isotherm.

Conversely, if by one means or another we know a characteristic curve of adsorption, we can use it to work back to the isotherm.

Polanyi makes the hypothesis that, for a given gas and solid, the characteristic curve of adsorption does not depend on the temperature. This hypothesis was verified by Titoff in the case of the adsorption of carbon dioxide by wood charcoal. Thus, on the basis of a characteristic curve of adsorption, we can obtain isotherms at other temperatures, and consequently, by using relation [6.54], calculate the equilibrium heat of adsorption $q_{\Delta\sigma}$. Figure 6.8(b) shows the characteristic curve of adsorption found by Titoff. Note the similarity in shape between the curve in Figure 6.8(b) and that given in Figure 6.4.

With the aim of determining the isotherms of adsorption of two different gases A and B on the same solid, Polanyi and Berenyi examined the relations between the characteristic curves of adsorption of the different gases on the same solid. On wood charcoal, they found the relation:

$$\frac{\Delta\varepsilon_A}{\Delta\varepsilon_B} = \left(\frac{a_A}{a_B}\right)^{1/2} \tag{6.128}$$

a_A and a_B denote the coefficients a of each of the two gases, A and B, in their respective van der Waals equations. If this expression is supposed to be satisfied for $\Delta\varepsilon \neq 0$ ($v \neq v_{max}$), we must have:

$$\ln\left(\frac{P^0}{P}\right)_A = \left(\frac{a_A}{a_B}\right)^{1/2} \ln\left(\frac{P^0}{P}\right)_B \tag{6.129}$$

This expression has been properly verified by Lindau.

This result led Dubinin to postulate that for all gases on a given solid, the characteristic curve of adsorption had the equation:

$$\Delta\varepsilon = \beta f(v) \qquad [6.130]$$

The function $f(v)$ is independent of the gas. The coefficient β is called the *affinity coefficient of the adsorbent.*

Experience has shown that for two gases A and B, the coefficients β have the same ratio as that existing between the molar volumes of the adsorbates in the adsorbed state, i.e. liquid, which is expressed by:

$$\frac{\beta_A}{\beta_B} = \frac{v^0_{(A)}}{v^0_{(B)}} \qquad [6.131]$$

Hence, we find:

$$\left(\frac{RT}{v^0_{(A)}}\ln\frac{P^0}{P}\right)_A = \left(\frac{RT}{v^0_{(B)}}\ln\frac{P^0}{P}\right)_B = f(v) \qquad [6.132]$$

Using this relation, it is possible to calculate an isotherm on the basis of another by way of the characteristic curves.

Certain criticisms have been leveled at this model. London noted that as the adsorption potentials vary in $1/r^3$, the values $\Delta\varepsilon$ quickly become negligible. This is undoubtedly compensated by the fact that the adsorbed layer itself has an adsorption potential. Lewis, for his part, proposed to replace the pressures by the fugacities, arguing that the proximity between the adsorbed phase and the liquid phase leads to not-insignificant interactions.

Ultimately, and paradoxically, the success of this model stems from the fact that it does not give an isotherm equation, but instead requires us, in order to find such an equation, to have at least one experimental curve at our disposal. This knowledge enables us, in view of the different observations, to find the isotherm of adsorption of the same gas on different solids at different temperatures.

6.6. Adsorption on porous substances

In the models we have discussed up until now, the gas was considered to be adsorbed on the whole of the surface of the solid, which was fully accessible. All points on that surface had equivalent (if not identical) properties. Thus, these models are completely legitimate to represent adsorption on:

– the surfaces of non-porous solids;

– solids exhibiting pores so small that the molecule of gas is not sterically able to penetrate them;

– solids whose pores have a sufficient value so that all the effects linked to the radii of curvature of those pores are negligible. The surface of those pores, then, presents no difference with the rest of the surface.

We shall now turn our attention to solids exhibiting medium porosity, whose adsorption is characterized by Type IV or V isotherms (see Figure 6.1).

Type IV isotherms present a sharply ascending part for high relative pressures, which closely follows the part represented by the B.E.T. isotherm, for instance (Figure 6.9). The quantities adsorbed on saturation become very great.

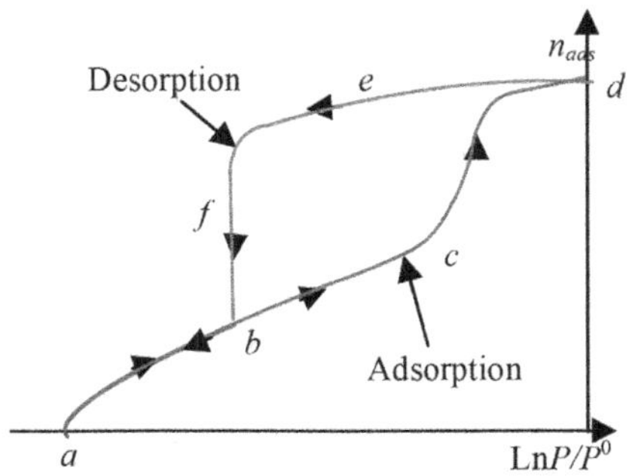

Figure 6.9. *Hysteresis of the Type IV isotherm*

In addition, for this type of isotherm, it has been observed that the curve plotted by decreasing the relative pressure (desorption curve) was not identical to that obtained by increasing the relative pressure (adsorption curve). We see an effect of hysteresis between the two parts of the curve (see Figure 6.9).

These two peculiarities of Type IV and V isotherms have been explained by including the phenomenon of capillary condensation in the pores of the adsorbent solid.

To present this explanation, in the following we shall consider a simple specific case which leads to a strictly-vertical ascendant part of the isotherm of adsorption (Figure 6.10).

6.6.1. *Process of pore filling*

Consider a solid whose pores are all identical to cylinders of the same radius r and open at both ends.

Starting from zero, when we increase the pressure of the gas, it is adsorbed to the whole surface, including the internal surface of the pores. In this step, we see multilayer adsorption far from saturation – i.e. which is accurately represented by the B.E.T. isotherm.

When the layer inside of the cylinders is sufficiently thick – a thickness estimated for a relative pressure of 0.35 – to have a surface tension to speak of, the cylindrical film has a saturating vapor pressure equal to that given by a cylindrical meniscus whose Kelvin radius is equal to the diameter of the cylinder and is given by relation [5.26]. The ratio of the pressures at equilibrium is then given by relation [4.31] (Figure 6.10(a)). We can show that this film becomes unstable in relation to a certain surface containing the same quantity of material, and which has the property of having a mean radius of curvature less than the radius of the cylindrical pore (Figure 6.10(b)).

As the pore continues to fill, that surface grows, and its mean radius of curvature decreases (see Figure 6.10(c)). Filling continues until the pore is obstructed by a biconcave liquid lens (Figure 6.10(d)). The liquid is then limited by a spherical meniscus whose average radius is equal to the radius of the cylinder, r.

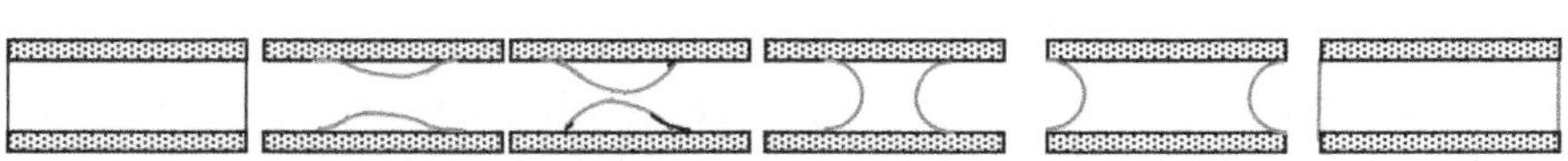

a) b) c) d) e) f)

Figure 6.10. *Capillary condensation in a cylindrical pore*

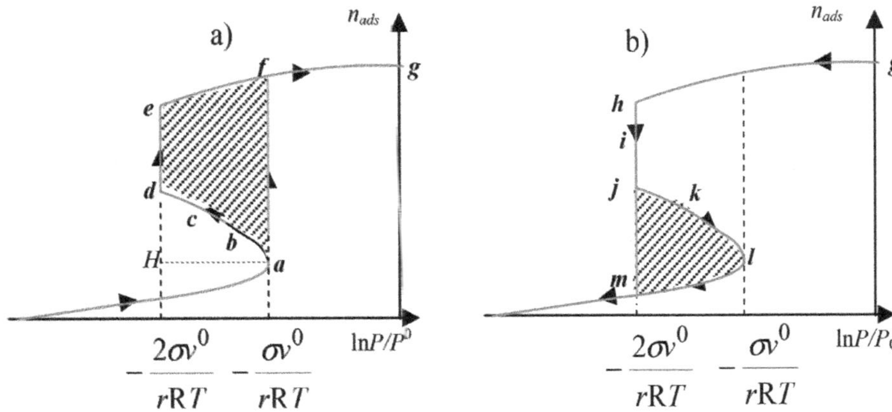

Figure 6.11. *a) Adsorption; b) desorption – with capillary condensation*

Filling continues (Figure 6.10(e)) as long as the radius of the meniscus is equal to the radius of the cylinder.

If we continue filling, the radius of the meniscus increases until it becomes infinite (Figure 6.10(g)).

6.6.2. *Shape of the adsorption curve*

On a diagram (Figure 6.11(a)), let us plot the quantity adsorbed as a function of the logarithm of the pressure ratio. On the abscissa axis, we place the two noteworthy values $-\dfrac{\sigma v^0}{rRT}$ and $-\dfrac{2\sigma v^0}{rRT}$, which correspond to the values of the pressure at equilibrium of a cylindrical meniscus and a spherical meniscus, both concave, with a Kelvin radius equal to the radius of the pore.

Figures 6.10(a)–(f) correspond respectively to points (a)–(f) in Figure 6.11(a).

We shall now plot the isotherm by setting different increasing values of the ratio P/P_0 (Figure 6.11(a)). Once we reach the point a, the experimental conditions will be out of equilibrium (which would be at H, equilibrium of a cylindrical meniscus) and the system will evolve spontaneously so as to re-establish its state of equilibrium – i.e. the figurative point will shift directly in terms of f irreversibly.

If the pores are not identical, it is easy to show that this results, on the isotherm, in a non-vertical ascending part, whose limit abscissa values depend on the limit radii of the pores.

6.6.3. *Shape of the evaporation curve, phenomenon of hysteresis*

As we saw earlier, it can be shown experimentally that the isotherm obtained by decreasing the pressure is not identical to the isotherm of adsorption.

This phenomenon of hysteresis can easily be explained, which we did using a simple example of a solid containing identical cylindrical pores with radius r. Figure 6.12(g) shows the state of the pores, which are full to begin with.

When the pressure is decreased, reversible evaporation takes place (see Figures 6.12(h), 6.12(i) and 6.12(j)) until the two spherical menisci of liquid are tangential to one another (Figure 6.12(j)). At that moment, the particular surface reforms (Figure 6.12(k)) and we again see the cylindrical meniscus. Figure 6.11(b) shows the corresponding plot of the desorption curve. Once we reach point j, the system, where the pressure is fixed, spontaneously and irreversibly evolves at m.

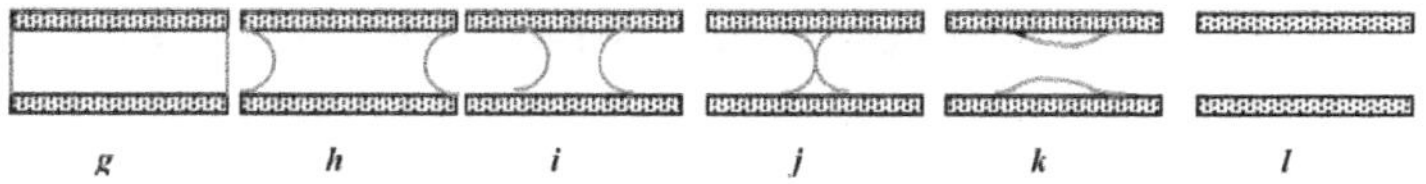

Figure 6.12. *Evaporation in a cylindrical pore*

Finally, the whole of the two plots appears in the form shown in Figure 6.13. The two verticals are such that:

$$\left(\frac{P}{P^0}\right)_d^2 = \left(\frac{P}{P^0}\right)_a \qquad [6.133]$$

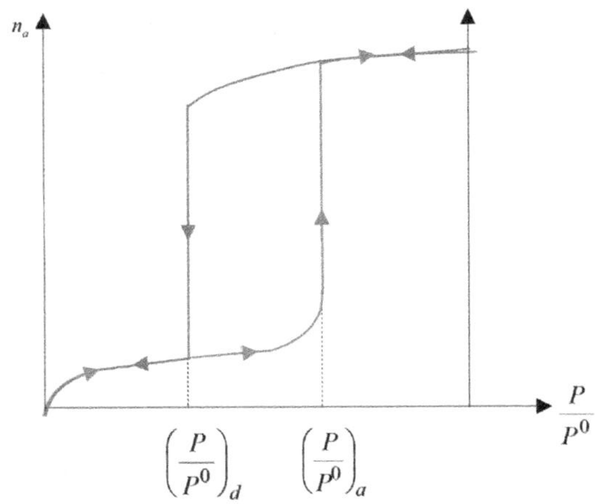

Figure 6.13. *Theoretical adsorption and desorption isotherms for identical cylindrical pores*

Hence, the classic process of multilayer adsorption followed by capillary condensation account for the isotherms of Type IV and V.

It is also worth noting that the vertical part of the adsorption does not correspond to a state of equilibrium, unlike the vertical part due to desorption. For this reason, it is the desorption curve that is used to determine the radii of the pores.

If the pores are not uniform, it can be shown that the desorption part remains practically vertical.

6.6.4. *Relationship between the shape of the pores and the hysteresis loop*

Barrer, de Boer and their collaborators studied the influence of pore shape on the shape of the hysteresis loop. Thus, de Boer distinguishes five types of hysteresis loops, which he calls A, B, C, D and E. As types A, B and E occur most commonly, we shall limit our discussion here to those three types, whose isotherms are shown in Figure 6.15.

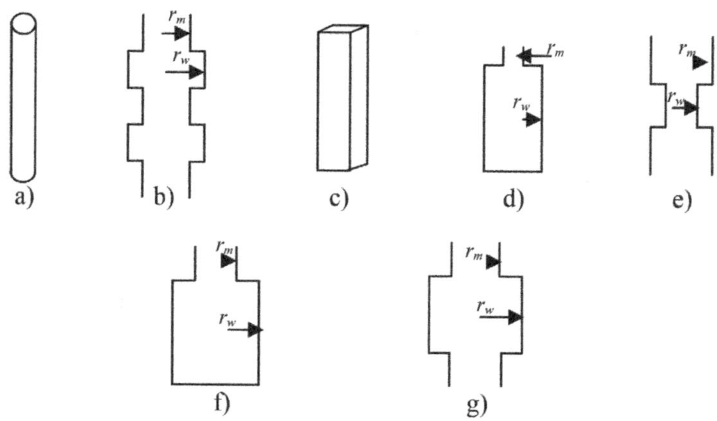

Figure 6.14. *Shapes of pores leading to type-A hysteresis loops*

Type A is found in the following cases (Figure 6.14):

– cylindrical capillary tubes open at both ends (Figure 6.14(a));

– expanded cylindrical capillary tubes with a small radius of aperture and with the condition on the radii $r_m < r_w < 2r_m$ (Figure 6.14(b));

– rectangular-section capillaries, open at both ends (Figure 6.14(c));

– cylindrical *ink-bottle* pores, with the conditions $r_m < r_w < 2r_m$ (Figure 6.14(d));

– cylindrical capillary tubes with a narrower part, where $r_m < r_w < 2r_m$ (Figure 6.14(e));

– cylindrical *ink-bottle* pores with the condition $2r_m < r_w$ (Figure 6.14(f));

– cylindrical *ink-bottle* pores open at both ends, where $r_w > 2r_m$ (Figure 6.14(g)).

Type B is found in the following two cases:

– very wide capillaries (radius greater than 50 nm);

– parallel planes at a certain distance from one another.

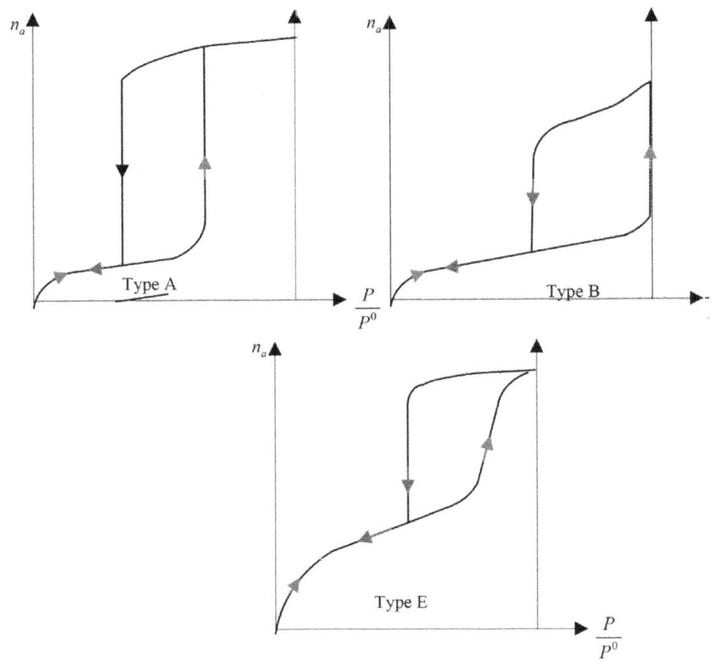

Figure 6.15. *Shapes of hysteresis loops for types A, B and E of the pore shapes according to De Boer's classification*

Type E strongly resembles type A; it can be attributed to open pores of radius r_m having spherical parts with radius r_w greater than $2r_m$.

If there is a distribution of the minimum radii r_m, the desorption branch will no longer be absolutely vertical.

In all cases where the pore does not have a uniform radius, the desorption branch corresponds to the radius of aperture of the pore (at the ends).

The shapes of the hysteresis loops are used to determine the pore radius distributions (see the Appendix).

Thus, we have used models to illustrate monolayer adsorption which leads to Type I isotherms, multilayer adsorption leading to Type III and Type IV isotherms, and multilayer adsorption and capillary condensation for Type V and VI isotherms in Brunauer's classification system.

7

Chemical Adsorption of Gases by Solids

Physical adsorption (see Chapter 6) involves physical bonds between molecules. We shall now describe a different type of adsorption, known as chemical adsorption, where gas molecules are linked to a solid by chemical bonds (molecular orbitals).

7.1. Chemical force between gas and solid surfaces

If the gas atom has single electrons, or if the dissociation of the gaseous molecule creates atoms with single electrons, or even if the gas molecule contains a multiple bond liable to be broken, there is the possibility of a chemical bond being formed between the gas and the solid. We shall distinguish between two types of solids: metals and semiconductors.

7.1.1. *Chemical adsorption on metals*

By magnetic studies, the main results have been obtained for the adsorption of gases on nickel.

With hydrogen, it has been shown that, at a constant temperature, the magnetization of the nickel decreases in a linear fashion with the amount of hydrogen fixed, and corresponds to two electron holes in the d band per molecule of hydrogen. This result proves that the adsorption of a molecule of hydrogen takes place with two bonds, rendering a dissociative chemical adsorption of hydrogen to nickel likely, meaning that the gas molecule is split into two atoms when bonded with the metal. The same result is found with nitrogen molecules.

The adsorption of a molecule of ethylene gives the same magnetic result as that of a hydrogen molecule, which appears to prove that the chemical adsorption of a molecule of ethylene takes place by means of its π electrons. At 100°C, the number of electrons transferred would be around 6, implying the dissociation of the ethylene molecule on adsorption.

The adsorption of a molecule of benzene fills 6 electron holes in band d, from which we deduce that the bond is made by the π electrons in the benzene ring.

It has also been shown that ethane is fixed by 6 electrons, which involves the breaking of three bonds in the gas molecule. By contrast, carbon monoxide is adsorbed without dissociation of the molecule.

Chemical adsorption of most gases results in a variation in the potential for electron extraction of the metal $\Delta\phi$. In the majority of cases, adsorption gives a negative film; however, ethylene, acetylene and vapors of alkali metals create positive films on the nickel. These results can be explained by the relative position of the Fermi level of the metal and the bond energy or the electron affinity of the gas. Indeed, if the ionization potential of the gas E_i is less than the metal's electron extraction potential, this means that the energy level of the electron in the adsorbate in the neutral state is higher than the Fermi level of the metal, so we have a polarized bond G^+M^- (Figure 7.1(a)).

If the electron affinity E_A of the adsorbate is greater than the work needed to extract the electrons from the metal, this means that there is an energy level of the electrons in the ionized gas in the form G^- which is lower than the Fermi level of the metal and we obtain a bond G^-M^+ (Figure 7.1(b)).

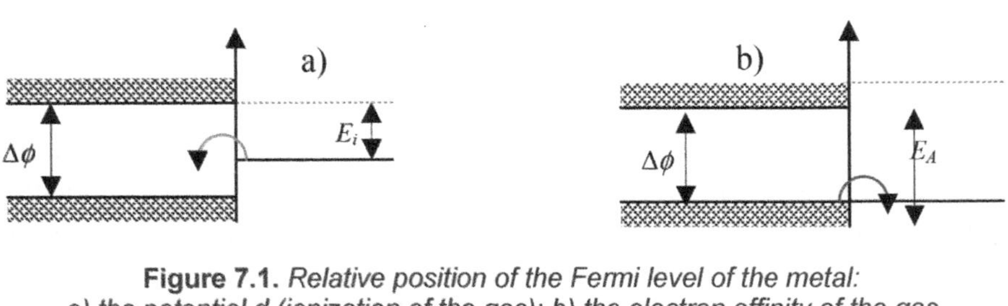

Figure 7.1. *Relative position of the Fermi level of the metal: a) the potential d (ionization of the gas); b) the electron affinity of the gas*

In all cases, if we adopt Helmholtz's hypothesis that the adsorbed layer forms a flat capacitor, of thickness d equal to the distance between the adsorbed particle and the support, and potential difference $\Delta\phi$, according to the classic relations of electrostatics, we have:

$$\Delta\varphi = \frac{Qe}{c} \quad [7.1]$$

Q is the adsorbed quantity, proportional to the degree of coverage θ, e is the elementary charge and c is the capacity of the capacitor with surface area S, which obeys the relation:

$$c = \frac{\varepsilon S}{d} \quad [7.2]$$

If μ denotes the dipole moment of the bond between the molecule and the metal, we deduce:

$$\Delta\varphi = \frac{\mu S\theta}{\varepsilon} \quad [7.3]$$

Thus, we see that the absolute value of the extraction potential varies in a linear fashion with the degree of coverage. Thus, in the case of a negative layer, if we ignore the energies of interaction between adsorbed particles, the enthalpy of adsorption would be:

$$\Delta_a H = E_A - \varphi = \Delta_a H^0 - b\theta \quad [7.4]$$

Thus, without bringing into play the concept of a heterogeneous surface, we can explain affine variations in the heat of adsorption with the degree of coverage.

Thus, by measuring the potential difference $\Delta\phi$, we are able to find the dipole moment of the bond, and therefore its ionic nature.

NOTE.– The use of relation [7.2] can be prevented if the distance between the solid and the linked molecule is of the same order of magnitude as that which separates the charges in the molecule. More sophisticated models have got around this difficulty and yielded qualitatively-similar results.

7.1.2. *Chemical adsorption on semiconductors*

Studies of chemical adsorption on semiconductors have shown that the electrical conductivity of the solid varies with the adsorbed quantity in one direction or the other. Also, at saturation, the covered fraction is often much less than the whole.

Three approaches have been taken to the modeling of the adsorption of gas on semiconductive solids: the concrete model method, the band diagram method and the valence line method. These methods – particularly the first two – must not be viewed as different models, but rather as different approaches to the same phenomenon.

7.1.2.1. *Concrete model method*

Consider the adsorption of a gas G to an ionic semiconductor formed of cations C^+ and anions A^-. Wolkenstein showed that it was necessary to take account of two types of bond between the gas and the solid: the *weak bond* and the *strong bond*.

In the so-called *weak* bond, the set formed by the adsorbed gas molecule G and its adsorption site undergoes no electrical alteration. The bond is formed by a single electron, similar to the bond in the molecule H_2^+. We shall represent such a bond by the symbol GL, with L denoting the lattice. A dipolar electrical moment occurs because of the movement, to a greater or lesser extent, of the electron clouds of the gas and of the support particle. The support particle may be either a cation or an anion in the lattice (see Figures 7.2).

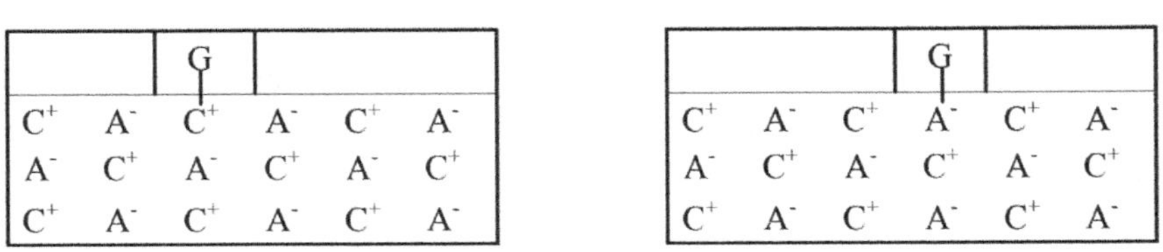

Figure 7.2. *Weak adsorption with only one electron*

In the so-called *strong* bond, the fixed particle keeps either an electron or a free electron hole close to it, and thus behaves like a charged particle. We can then see that the electrical conductivity can be modified. If an electron is retained, the bond formed will be an n bond (written as GeL); if an electron hole is retained, then it will be a p bond (GpL). We can easily see that the adsorption of an electron acceptor gas will increase the conductivity by fixing to a semiconductor p, whereas that conductivity would be decreased on a semiconductor of type n. The opposite result is obtained by the adsorption of an electron donor gas.

The strong bond leads to the formation of a double electrical layer, positive or negative on the outside and of the opposite sign on the inside. This double layer renders it more difficult to extract electrons from the solid, which necessarily decreases the enthalpy of adsorption. In the same way, a calculation similar to that put forward by Helmholtz for metals gives us relation [7.4]. Thus, we find the linear decrease of the enthalpy of adsorption.

According to Wolkenstein, only weakly-bonded gas molecules are liable to desorb in the same state as they were in before adsorption.

7.1.2.2. *Band diagram model*

An ionic semiconductor can be represented by a band diagram showing two energy bands: a valence band for lower energy levels and a conduction band for higher energies. These two bands are separated by the forbidden band (gap), whose "width" is a characteristic of the semiconductor. In the forbidden band, a few levels are authorized by the presence in the semiconductor of point defects that are electron donors or acceptors (Figure 7.3).

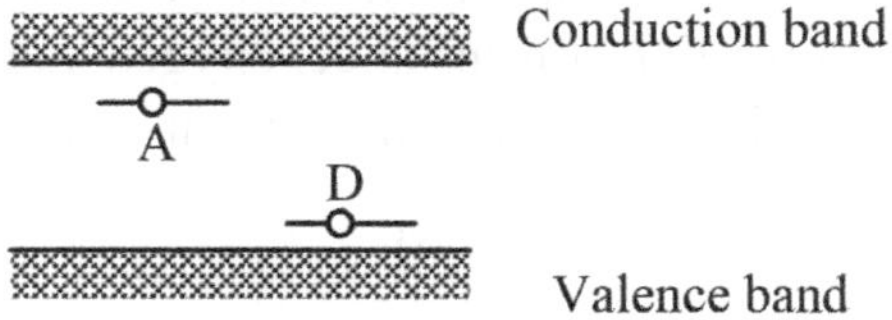

Figure 7.3. *Energy position of a donor and an acceptor*

It has been shown that a foreign particle G chemically adsorbed to the surface by a weak bond has its "image" at level A or at level D, depending on whether G is an electron acceptor or donor. The appearance of an electron on level A indicates the passage from the weak bond to the strong bond, and this may take place either by the falling to level A of electrons from the conduction band (e.g. oxygen on ZnO), or by the raising to A of an electron from the valence band (e.g. oxygen on Cu_2O). The same reasoning can be applied for level D, near to the valence band. The creation of the strong bond causes the appearance of a surface electric charge. This results in a curvature of the band diagram in the vicinity of the surface, which leads to an alteration of the Fermi level, and therefore a limitation of the adsorbed quantity. Figures 7.4(a) and 7.4(b) respectively show the shape of the bands in the adsorption of an electron acceptor to a semiconductor n and of an electron donor to a semiconductor p.

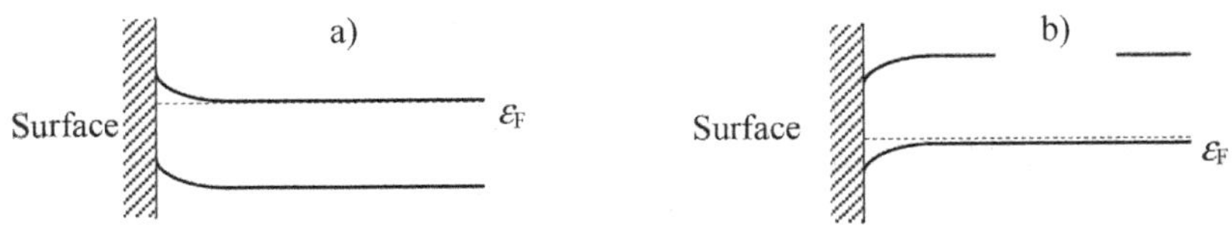

Figure 7.4. *Shape of the bands in a semiconductor on the adsorption: a) of an acceptor gas to type-n; b) of a donor to type-p*

7.1.2.3. *Valence lines model*

In this model, we consider that the free electrons and electron holes in the semiconductors constitute free valences on the surface of the solid. Figures 7.5(a), 7.5(b) and 7.5(c) respectively show the valence arrangement of a weak bond, a strong bond n and a strong bond p. The localization is not specified, other than the fact that the electron defect is associated with an ionic point defect, and we find, for instance, oxygen vacancies as the site of adsorption of an oxide n liable to accommodate an electron donor gas.

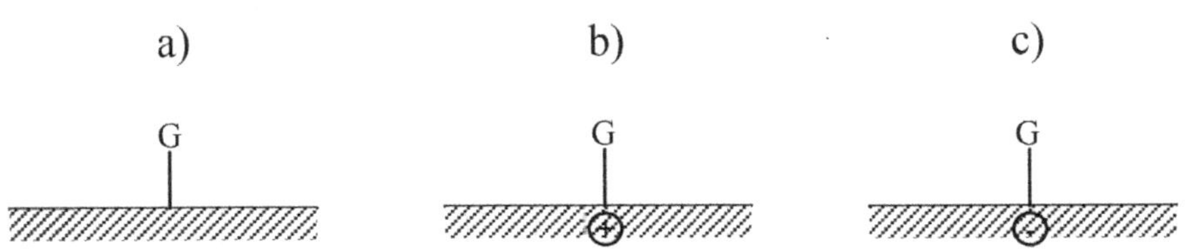

Figure 7.5. *Different types of valence lines on a semiconductor*

7.2. Physical adsorption and chemical adsorption

Let us return to the comparison between the potential energy of the adsorbed molecule and the thermal (kinetic) energy k_BT. We have already used this comparison for the instantaneous potential energy, which enabled us to differentiate mobile physical adsorption and localized adsorption (see section 6.4.1).

We shall now compare the mean potential energy of a gas molecule approaching a solid using thermal energy. We shall base our discussion on the example of the adsorption of nitrogen to nickel, and examine three cases:

– the mean potential energy is less than the heat energy, in which case we speak of collision rather than adsorption, and the phenomenon can simply be modeled as the specular backscattering of the gas molecules onto the surface of the solid. We say that the *accommodation coefficient*, which is defined as the probability of a gas molecule of being adsorbed on contact, is nearly zero. This coefficient will be nearer to 1 when the interaction energy is close to the kinetic energy of the gas. The molecules spend only a very short period of time – around 10^{-13}s – in interaction with the solid wall (this is the period of oscillation at the minimum of the potential trough);

– the mean potential energy is around a few k_BT, in which case we have physical adsorption. In this case, the accommodation coefficient is near to 1, and the residence time in the vicinity of the surface is of the order of 10^{-7}s at a temperature of 300 K. The curve giving the potential energy as a function of the distance from the wall reaches a minimum at equilibrium (Figure 7.6), with an attraction in $1/r^3$ (see section 6.2);

– the mean potential is greater than 10 or 20 times the kinetic energy k_BT. We then have chemical adsorption. The adsorption is attended by the breaking of the N-N chemical bond, which produces a curve showing an energy of activation of adsorption, E_a, before reaching the position of equilibrium of chemical adsorption (Figure 7.6). The accommodation coefficient will be 1 and the residence time around 10^{+17}s at a temperature of 300 K.

Table 7.1 presents the details for the three cases envisaged here.

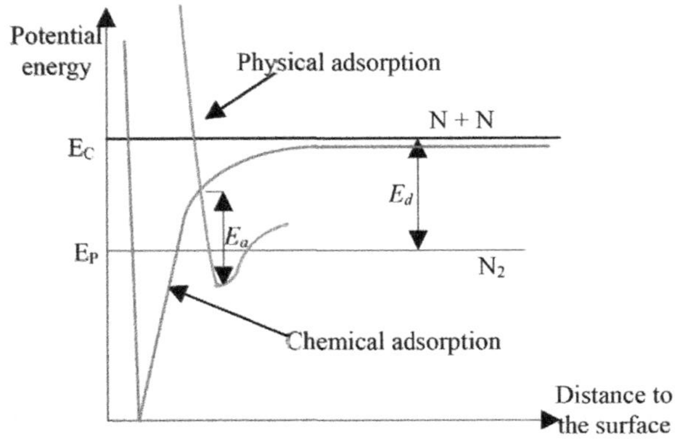

Figure 7.6. *Physical adsorption and chemical adsorption*

Energy of interaction (kJ/mole)	Residence time (s)	Accommodation coefficient	Phenomenon
0.4	10^{-13}	0	Collision
40	10^{-7}	1	Physical adsorption
180	10^{+17}	1	Chemical adsorption

Table 7.1. *Comparison of the differing behaviors of a gas molecule as it approaches a solid surface*

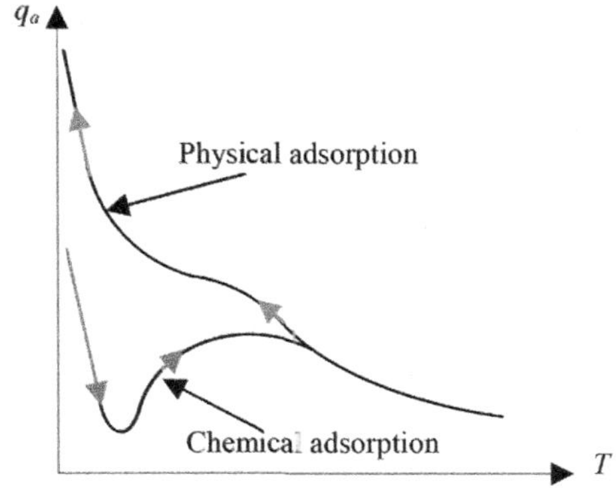

Figure 7.7. *Transition from physical adsorption to chemical adsorption*

The existence of an activation energy for the process of chemical adsorption means that this process tends to take place at a high temperature, whereas physical adsorption, which is a non-activated process, can occur at a lower temperature. It is possible for the same gas–solid couple to be involved in both phenomena in succession as the temperature increases, as is shown by Figure 7.7. This figure illustrates that, when the temperature rises, the absorbed quantity (isobar of adsorption) passes from a characteristic curve of physical adsorption to a characteristic curve of chemical adsorption.

7.3. Isotherms of adsorption and experimental results

As the bond between the gas and the surface of the solid is chemical in nature, this often results in the existence of a monolayer of gas at the maximum. That same bond means that we have localized adsorption.

As in the case of physical adsorption in a monolayer, we very often choose the fraction of coverage θ (or degree of filling), at equilibrium, of the surface as the value linked to the adsorbed quantity.

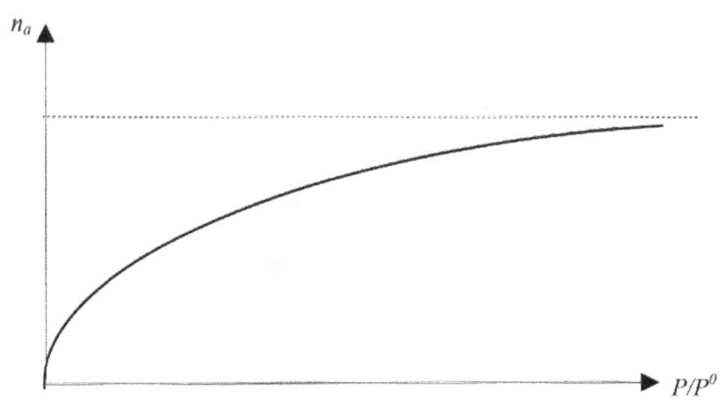

Figure 7.8. *Isotherm of chemical adsorption*

Experience shows us that chemical adsorption is also a divariant phenomenon. The quantity fixed at equilibrium is a function of the gas pressure and of the temperature.

Figure 7.8 shows the typical shape of an isotherm of chemical adsorption. Note a saturation effect at high pressures.

The difficulty of obtaining isotherms experimentally usually lies in the achievement of low gas pressures that are perfectly controlled, which cannot be obtained directly, but instead are established on the basis of a chemical equilibrium which determines the gas pressure (for example the fixation of very low pressures of pure oxygen because of the tension of dissociation of an oxide).

The phenomenon of chemical adsorption is exothermic, which results in a decrease of the quantity adsorbed at equilibrium as the temperature increases, which has been verified experimentally.

It is very helpful to be able to measure the heat of adsorption. This can be done directly by calorimetry, or indirectly on the basis of the isotherms obtained at different temperatures. Experimentally, we see that the heat of adsorption frequently varies with the fraction of coverage (Figure 7.9). The values given by the tables, therefore, are usually the initial heats of adsorption (with zero coverage).

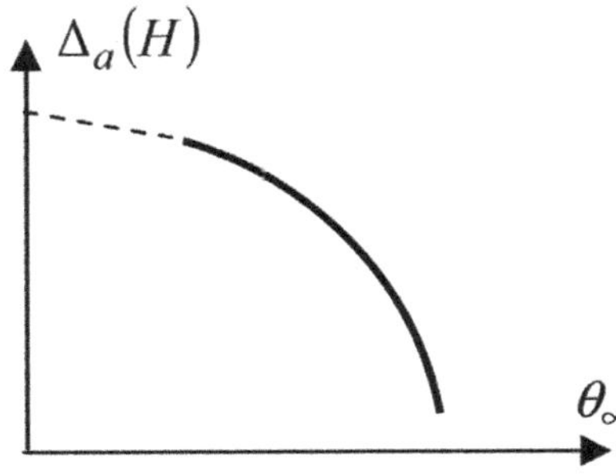

Figure 7.9. *Variation in enthalpy of chemisorption with degree of coverage*

7.4. Langmuir's model of equilibrium of chemical adsorption

The above characteristics, which are comparable to those of localized monolayer physical adsorption (see section 6.4.4), mean that Langmuir's is the fundamental model for equilibria of chemical adsorption.

Remember that in the conditions of that model, the equilibrium of chemical adsorption is written:

$$\mathrm{G+s=G-s} \qquad [7R.1]$$

where the species s and G–s are two components of the same solution.

The law of mass action, applied for the equilibrium of the previous reaction, is written:

$$\frac{\gamma_{\mathrm{G-s}}\theta}{P\gamma_{\mathrm{s}}(1-\theta)} = K \qquad [7.5]$$

where $\gamma_{\mathrm{G-s}}$ and γ_{s} respectively denote the activity coefficients of the components of the solution: s and G–s.

As K is an equilibrium constant, it obeys van 't Hoff's law with changing temperature, so:

$$K = K^0 \exp\left(- \frac{\Delta_a H^0}{\mathrm{R}T} \right) \qquad [7.6]$$

The activity coefficients are not independent: at constant total pressure and temperature, they are linked by the equation derived from the Gibbs–Duhem equation:

$$\theta d \ln \gamma_{\mathrm{G-s}} + (1 - \theta) d \ln \gamma_{\mathrm{s}} = 0 \qquad [7.7]$$

If we examine the particular case of a surface which would constitute a perfect solution of free and occupied sites, the activity coefficients are both equal to 1, and obviously, we have Langmuir's isotherm:

$$\theta = \frac{KP}{1+KP} \qquad [7.8]$$

In view of the variations in the equilibrium constant with temperature, given by relation [7.6], Langmuir's equation takes the form:

$$\theta = \frac{K^0 P \exp - \left(\frac{\Delta_a H^0}{\mathrm{R}T} \right)}{1 + K^0 P \exp - \left(\frac{\Delta_a H^0}{\mathrm{R}T} \right)} \qquad [7.9]$$

This Langmuir equation is obviously identical to that found in physical adsorption; only the orders of magnitude of the enthalpies are different – around 20 kJ/mole in physical adsorption, as opposed to around 100 kJ/mole in chemical adsorption.

7.5. Dissociative adsorption and Langmuir's model

It is difficult to imagine how stable, saturated gas molecules (H_2, O_2, etc.) could form a chemical bond with an atom at the surface of the solid. To explain this bond, as we saw in section 7.1, we are led to accept the hypothesis, in numerous cases, of dissociation of the gaseous molecule accompanying the adsorption.

Thus, consider a gas G_i with atomicity i which dissociates on adsorption, according to:

$$\mathrm{G}_i = i\mathrm{G} \tag{7R.2}$$

The equilibrium of fixation is written as:

$$\frac{1}{i}\mathrm{G}_i + \mathrm{s} = \mathrm{G-s} \tag{7.10}$$

The same reasoning as in the previous section yields:

$$\frac{\gamma_{\mathrm{G-s}}\theta}{P^{1/i}\gamma_{\mathrm{s}}(1-\theta)} = K \tag{7.11}$$

which, in the case of a perfect solution, gives us:

$$\frac{\theta}{P^{1/i}(1-\theta)} = K \tag{7.12}$$

This is Langmuir's new equation in the case of dissociation of the gas on adsorption. The curve obtained is of the same shape as that represented in Figure 7.8.

It may be that the different fragments of a gas AB do not attach to the same sites, so for instance, for a gas dissociating into two entities A and B, the equilibrium is written:

$$AB + s_A + s_B = A - s_A + B - s_B \quad [7R.3]$$

The equilibrium condition gives us:

$$\frac{\theta_A \theta_B}{P(1-\theta_A)(1-\theta_B)} = K \quad [7.13]$$

If, in addition, the fractions of coverage of the two entities are identical, we obtain:

$$\frac{\theta_A^2}{P(1-\theta_A)^2} = K \quad [7.14]$$

This relation is of the same form as equation [7.12] for $i = 2$.

7.6. Chemical adsorption of mixtures of gases in Langmuir's model

Consider the adsorption of a mixture of two gases A and B, which are adsorbed to the same surface. Then, two scenarios may arise:

– either there are sites specific to A and sites specific to B, in which case the two adsorption reactions are independent, and the equations of the isotherms are given by relations [7.8] or [7.14];

– or the adsorption sites of the solid are the object of competition between A and B. Thus, let θ_A and θ_B denote the fractions of the surface at equilibrium respectively covered by A and B, and let P_A and P_B by the partial pressures of each of the two gases. The equilibria of adsorption will be:

$$A+s=A-s \quad [7R.4]$$

and

$$B+s=B-s \quad [7R.5]$$

The molar fractions of the species will, respectively, be:

$$x_{A-s} = \theta_A \qquad [7.15a]$$

$$x_{B-s} = \theta_B \qquad [7.15b]$$

and:

$$x_s = 1 - \theta_A - \theta_B \qquad [7.15c]$$

Thus, let us apply the law of mass action to each of the two equilibria; the isotherms become:

$$\frac{\gamma_{A-s}\theta_A}{P_A\gamma_s(1 - \theta_A - \theta_B)} = K_A \qquad [7.16a]$$

and

$$\frac{\gamma_{B-s}\theta_B}{P_B\gamma_s(1 - \theta_A - \theta_B)} = K_B \qquad [7.16b]$$

In the context of Langmuir's hypothesis of the perfect solution, we obtain:

$$\frac{\theta_A}{P_A(1 - \theta_A - \theta_B)} = K_A \qquad [7.17a]$$

and

$$\frac{\theta_B}{P_B(1 - \theta_A - \theta_B)} = K_B \qquad [7.17b]$$

which gives us, for the coverage of each species:

$$\theta_A = \frac{K_A P_A}{1 + K_A P_A + K_B P_B} \qquad [7.18a]$$

and

$$\theta_B = \frac{K_B P_B}{1 + K_A P_A + K_B P_B} \qquad [7.18b]$$

We can see that the ratio between the fractions of coverage is proportional to the ratio between the partial pressures.

$$\frac{\theta_A}{\theta_B} = \frac{K_a}{K_B} \cdot \frac{P_a}{P_B} \qquad [7.19]$$

These results can easily be generalized to the adsorption of more complex mixtures, and we can show that for the gas A, belonging to a mixture, we have:

$$\theta_A = \frac{K_A P_A}{1 + \sum_j K_j P_j} \qquad [7.20]$$

where j represents all of the gases present in the mixture, which are adsorbed to the same sites as the gas A.

NOTE.– If the gases are dissociated and the sites competitive, in repeating the method in the previous section, we need to replace the partial pressures with those pressures assigned to the power $1/i_j$, where i_j denotes the atomicity of the gas j.

7.7. "Non-Langmuirian" isotherms of adsorption

Langmuir's expression (and its derivatives) is considered to be verified by experience, as the accuracy of the measurements was so poor. Today, we know that this expression represents a reasonably good approximation. However, it is very useful, because all expressions established more recently derive from Langmuir's; in addition, it is a simple analytical expression which, most of the time, is sufficient when adsorption is one of the phenomena, along with others, occurring in a heterogeneous reaction, for example.

Two methods have been used to obtain other representations of the isotherm of adsorption.

With the first method, we consider that not all the adsorption sites are equivalent. The distribution of sites, which may be continuous or discontinuous, is characterized by their heats of adsorption. It is not possible to detect whether that heterogeneity of the sites pre-exists the adsorption on the surface of the solid or if it is created as the surface becomes filled following the interaction between the affixed molecules (or both reasons). The two origins obviously lead to the same result.

In this method, we apply Langmuir's relation [7.9] to sites in category i, characterized by an enthalpy $\Delta_a H_i^0$, and add across all the categories of sites.

Various energy distributions have been put forward:

– suppose we have a single type of sites; we then obviously find Langmuir's relation again;

– suppose we have an exponential distribution of sites:

$$n_i = n_0 \exp\left(-\frac{\Delta H_i^0}{\Delta H_0^0}\right) \qquad [7.21]$$

by summing together across all the values of θ_i, then for small values of θ, we find:

$$\ln\theta = \frac{RT}{\Delta H_0^0}\ln P + \text{Constant} \qquad [7.22]$$

so:

$$\theta = kP^n \qquad [7.23]$$

This is the expression of *Freundlich's isotherm*;

– other authors posit that the heat of adsorption is an affine function of the degree of coverage, meaning that the curve in Figure 7.9 is assimilated to a straight line (remember that such a curve was justified in section 7.1, by relation [7.4]). Thus, we posit:

$$\Delta H^0 = \Delta H_0^0(1 - \beta\theta) \qquad [7.24]$$

ΔH_0^0 represents the initial heat of adsorption. Substitute that relation back into expression [7.9], and let us set:

$$A = K^0 \exp\left[- \frac{\Delta H_0^0}{\mathrm{R}T} \right] \qquad [7.25]$$

We obtain:

$$\frac{\theta}{1-\theta} = AP \exp\left[\frac{\Delta H_0^0 \beta \theta_\infty}{\mathrm{R}T} \right] \qquad [7.26]$$

and expressed in logarithmic form, this gives us:

$$\ln P = \ln \frac{\theta}{1-\theta} - \ln A - \frac{\Delta H_0^0 \beta \theta_\infty}{\mathrm{R}T} \qquad [7.27]$$

In the range of average coverage (θ approximately 0.5), the first term on the right-hand side can be discounted, and we find the isotherm put forward by Temkin:

$$\theta = - \frac{\mathrm{R}T}{\Delta H_0^0 \beta} \ln AP \qquad [7.28]$$

The second method for obtaining "non-Langmuirian" expressions is based on the general relation of the model [7.5], with which we associate a model of a non-perfect solution, which will express the activity coefficients as a function of the molar fractions – i.e. as a function of the degree of coverage. For instance, if we take the model of strictly-regular solutions, the activity coefficients are of the form:

$$\ln \gamma_{\mathrm{G-s}} = \frac{w}{\mathrm{R}T}(1 - \theta)^2 \qquad [7.29a]$$

and

$$\ln \gamma_s = \frac{w}{RT}\theta^2 \qquad [7.29b]$$

By substituting back into relation [7.5], we find:

$$\frac{\theta}{1-\theta} = KP \exp\left[\frac{w(2\theta - 1)}{RT}\right] \qquad [7.30]$$

which is of the same form as *Tempkin's isotherm.*

This second method for introducing heterogeneity by taking into account the interactions between adsorbed molecules and free sites (by the activity coefficients) is richer than the first method. Indeed, the models of solutions often have a physical basis, but the energy distributions in the first method are often set *a priori*. In addition, this second method is very well suited to all the different types of adsorption encountered (dissociative, mixture of gases, etc.).

Appendix

Applications of Physical Adsorption to the Study of the Area and Porosity of Solids

Physical adsorption is at the root of two measurements characteristic of solids:

– the specific area;

– the pore radius distribution.

A.1. Determination of the specific area of a solid

The specific area is defined as the total external area accessible to a gas, per unit mass, usually in m^2/g. To measure it, we use the B.E.T. method.

A.1.1. *Specific area and capacity of a monolayer*

The capacity of a monolayer v_{mono} of a solid in relation to a gas is defined as being the volume of gas, expressed in normal conditions of pressure and temperature, which would be necessary to cover the surface of one gram of the solid with a complete monolayer on adsorption of the gas. This value is fictitious, because clearly, upon adsorption, the creation of upper layers begins before the monolayer is fully in place. When the quantity v_{mono} is reached, this means that there are sufficient molecules adsorbed in the different layers to completely cover the solid in a monolayer.

Let q_m represent the number of molecules of gas adsorbed per gram of solid to form a monolayer, and σ_{mol} the surface occupied by a molecule of gas on the surface of the solid, measured in square angström. Thus, for the specific area, we find:

$$\Sigma(m^2/g)=10^{-20}q_m\sigma_{mol}(\text{Ä}^2) \qquad [A.1]$$

However, the number q_m is linked to v_{mono} by the relation:

$$q_m=\frac{v_{mono}}{22400}6.02\times10^{23}=0.269v_{mono} \qquad [A.2]$$

This gives us the specific area:

$$\Sigma(m^2/g)=0.269v_{mono}(cm^3/gT,P,N)\sigma_{mol}(\text{Ä}^2) \qquad [A.3]$$

Note the particular units usually adopted in measuring the specific areas.

A.1.2. *Areas of the molecules*

In evaluating the area occupied on the solid by an adsorbed molecule, we suppose for simplicity's sake that these molecules are spherical and that the molecules on the surface of the solid present a compact hexagonal arrangement. If we let M denote the molar mass, ρ_L the specific mass of the adsorbed liquid product and N_a Avogadro's number, we deduce:

$$\sigma_{mol}=2\sqrt{3}10^{16}\left(\frac{M}{4\sqrt{2}N_a\rho_L}\right) \qquad [A.4]$$

Hence, for instance, we choose the value 16.27 Ä^2 for the area of the nitrogen molecule at the temperature of liquefaction of nitrogen, and 32.1 Ä^2 for the butane molecule.

A.1.3. *Measuring the capacity of a monolayer*

Various methods, based on physical adsorption, are used to evaluate the capacity of a monolayer and ultimately the specific area. The main two ones are:

– the B-point method, which is fast but not hugely accurate;

– the B.E.T. method, which takes longer to implement but yields more definite results.

A.1.3.1. *The B-point method*

The B-point method is essentially a fast and simple method to evaluate a specific area. Emmet and Brunauer, considering Type II isotherms (see Figure 6.1), believed that the monolayer capacity should correspond to a remarkable point on the isotherm. According to the authors, such an isotherm exhibits five remarkable points called A, B, C, D and E (see Figure A.1).

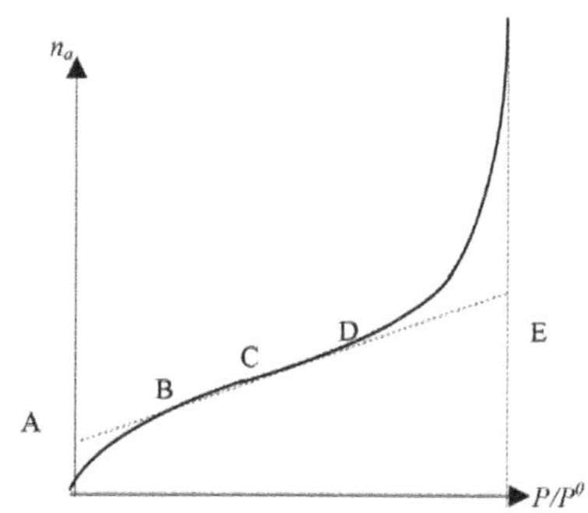

Figure A.1. *Isotherm of adsorption*

Point chosen	Measured error
A	7-28%
B	3-12%
C	5-17%
D	7-20%
E	11-27%

Table A.1. *Errors in terms of the specific areas obtained based on the remarkable points on the isotherms*

They measured the specific area of known solids, supposing that the five remarkable points corresponded successively to the volume of the monolayer. The differences obtained, for each of the points, with the true value of the specific area are given in Table A.1.

Note that point B is the most satisfactory. This result was confirmed by the same authors, noting that curve giving the heat of adsorption as a function of the adsorbed quantity often showed a maximum for the value corresponding to that of point B on the isotherm.

In addition, Halsey showed that this point B was where the affinity of the surface for the gas changed most quickly.

Ultimately, it is unsurprising that no remarkable point on the curve corresponds to the specific surface, given that the monolayer is simply a mental construction and that fixation in multiple layers begins long before the surface has been fully covered.

A.1.3.2. *The B.E.T. method*

The B.E.T. method is much more accurate than the B-point method, and can be considered a true analytical method for measuring the specific area.

Equation [6.117] enables us to calculate the capacity of the monolayer v_{mono} (see section 6.5.1). In general, experience yields a straight line for values of the ratio between the pressures of between 0.05 and 0.35. Thus, we can say that the B.E.T. model is correct in that range.

To take the measurements, it is advantageous to choose a gas which gives a high value of the parameter c. Indeed, for $V = v_{mono}$, we can write $\frac{P}{P^0} v_{mono} = \frac{Vc-1}{c-1}$. The higher the value of c, the smaller the ratio $\frac{P}{P^0}$ is. This is why nitrogen is often used at the temperature of liquid air.

In the case that $c >> 1$, and therefore $\frac{c-1}{c} \approx 1$, we obtain $\frac{1}{cv_{mono}} \approx 0$ and the equation of the B.E.T line becomes:

$$\frac{P}{V\left(P^0 - P\right)} = \frac{1}{v_{mono}} \frac{P}{P^0} \qquad \text{[A.5]}$$

Then, the line passes through the origin, and thus a single point is sufficient to measure the capacity of the monolayer v_{mono}.

For measuring very small specific areas (around 1 m^2/g), a small fraction of gas is adsorbed, so the quantity adsorbed cannot be known with precision. This difficulty can be overcome by working at temperatures where the vapor pressure of the adsorbed gas is sufficiently low to increase the value of the ratio between the pressures. For this purpose, it is common to use krypton at the temperature of liquid air; in such a case, the gas has a saturating vapor pressure of around 2 hPa. In these conditions, we can go far below 1 m^2/g, using samples of a few grams of solid.

In any case, the measurement of the specific area of a solid is one which does not yield a high degree of precision, and it would be illusory to envisage values with precision greater than 10%.

A.2. *Determination of the pore radii based on the isotherms of adsorption*

We have seen (section 6.6) that the pore radius is linked to the isotherms of physical adsorption with a hysteresis loop; hence the idea to use these isotherms to try to deduce a distribution of the radii of the open pores in a solid.

A.2.1. *Pore radius at equilibrium at a given pressure*

At a point on the desorption branch of the hysteresis loop, the liquid is condensed in spherical pores with a given Kelvin radius such that:

$$\ln\frac{P}{P^0} = -\frac{2\sigma v^0}{r_K \mathrm{R}T} \qquad [A.6]$$

This Kelvin radius is, in fact, the radius r of the pore less the thickness t of the adsorbed layer for the given pressure ratio P/P^0. Thus, we have:

$$r = r_K + t \qquad [A.7]$$

To determine r, then, we need to know the desorption curve of a gas to find r_K, and its adsorption curve to determine t.

For these measurements, it is usual to use the adsorption of nitrogen, and we shall see how to calculate the thickness of the adsorbed layer t.

A.2.2. Calculation of thickness t of the adsorbed layer

To determine t we can, as Linsen did, assimilate the adsorbed gas to the liquid. Let X denote the volume of liquid adsorbed in cm^3 per gram of solid, and S the surface area covered in m^2 per gram of solid. Immediately, we can write:

$$t(\text{Ä}) = \frac{X.10^{24}}{S.10^{20}} = \frac{X}{S}10^4 \quad [A.8]$$

However, if we assimilate the adsorbed gas to the liquid, and if V is the volume of the adsorbed gas in cm^3, we can write:

$$X = \frac{\text{mass of liquid}}{\text{density of the liquid}} = \frac{\text{mass of adsorbed gas}}{\text{density of the adsorbed gas}} = \frac{MV}{22400\rho} \quad [A.9]$$

By feeding this back into relation [A.8], we obtain:

$$\mathrm{t} = \frac{10^4 MV}{22400\rho S} \quad [A.10]$$

In the case that the adsorbed gas is nitrogen, we have the following values: $\rho = \frac{1}{12.4}\,\text{g/cm}^3$ and $\mathrm{M} = 28$, so:

$$t = 15.47\frac{V}{S} \quad [A.11]$$

De Boer accepts that the surface of the pores is practically equal to the specific surface measured by the B.E.T. method which, in the case of nitrogen, tells us:

$$S = S_{\text{BET}} = 16,27 \times 0.269 v_{\text{mono}} = 4.37 v_{\text{mono}}\,\text{m}^2/\text{g} \quad [A.12]$$

Hence, the thickness of the adsorbed layer:

$$\mathrm{t} = 3.54\frac{V}{v_{\text{mono}}} \quad [A.13]$$

In this expression, V is the volume adsorbed excluding the volume condensed in the pores.

Hence, in order to calculate the pore radius, it is helpful to know the function $t\left(\frac{P}{P^0}\right)$, which can be used to calculate the thickness t on the basis of the isotherm without having to first calculate the capacity of the monolayer v_{mono}.

Three methods can be used to determine the function $t\left(\frac{P}{P^0}\right)$.

A.2.2.1. *De Boer's method*

De Boer calculated the quantity $3.54\frac{V}{v_{\text{mono}}}$ for different solid samples. He plotted that quantity as a function of the ratio P/P^0.

For values of P/P^0 less than 0.75, he notes that the points relative to the different solid samples all lie on the same curve.

For values of the ratio P/P^0 slightly greater than 0.75, the curves relative to each sample separate slightly. The second column in Table A.2 gives the values obtained by de Boer.

P/P^0	t(Ä) de Boer	t(Ä) B.E.T.	t(Ä) F.H.H.
0.1	3.63	3.91	4.60
0.2	4.36	4.42	5.17
0.3	5.01	5.05	5.70
0.4	5.71	5.90	6.23
0.5	6 .50	7.08	6.83
0.6	7.36	8.92	7.61
0.7	8.57	11.9	7.89
0.8	10.6	17.8	9.98
0.9	15	35.4	12.92

Table A.2. *Compared values of the thicknesses of adsorbed layers as a function of the ratio between the pressures, obtained using different methods*

A.2.2.2. *B.E.T. method*

If we accept equation [6.116] for the isotherm, then we can deduce:

$$\frac{V}{v_{mono}}=\frac{c\dfrac{P}{P^0}}{\left(1-\dfrac{P}{P^0}\right)\left[1+(c-1)\dfrac{P}{P^0}\right]} \qquad [A.14]$$

However, for nitrogen, we have $c \gg 1$, and for a certain range of the pressure ratio $c\dfrac{P}{P^0}\gg 1$, the equation of the isotherm is reduced to:

$$\frac{V}{v_{\text{mono}}}=\frac{1}{1-\dfrac{P}{P^0}} \qquad [A.15]$$

This gives us the following value of the thickness of the adsorbed layer:

$$t=\frac{3.54}{1-\dfrac{P}{P^0}} \qquad [A.16]$$

The third column in Table A.2 gives the values of t thus obtained. Note that they match closely with those found by de Boer as long as the pressure ratio P/P^0 is less than 0.4 – in fact, in the domain of validity of the B.E.T. equation.

A.2.2.3. *Frenkel, Halsey and Hill (F.H.H.) method*

It has been remarked that for high relative pressures, the F.H.H. isotherm was far more satisfactory than the B.E.T. isotherm. We saw earlier (in section 6.5.2) that this isotherm is expressed in the form:

$$\mathrm{R}T\ln\frac{P}{P^0}=-\frac{a}{\Gamma^3}=-\frac{K}{\left(\dfrac{V}{v_{\text{mono}}}\right)^3} \qquad [A.17]$$

From this, the following ratio was derived:

$$\frac{V}{v_{\text{mono}}} = \left(\frac{K}{RT \ln \frac{P^0}{P}} \right)^{1/3} \qquad \text{[A.18]}$$

For nitrogen at 77 K, the proposed ratio $K/RT = 5$, so we have the following value of t:

$$t = 3.54 \left(\frac{5}{2.3 \log \frac{P^0}{P}} \right)^{1/3} \qquad \text{[A.19]}$$

The last column in Table A.2 shows that the values obtained match well with de Boer's results.

A.2.3. *Determination of the pore radius distribution in a solid*

Thus, consider a solid whose pores do not all have the same radius. We want to plot the curve giving the volume of the pores as a function of their radius, which is called the pore volume distribution curve. We suppose that the hysteresis loop for the adsorption of nitrogen to the solid in question is known.

A.2.3.1. *General method*

At pressure P_n, all the pores of radius $r < r_n$ are filled with liquid. Divide the abscissa axis of the isotherm (axis of relative pressures) into equal intervals. If the pressure is decreased from P_n to $P_{n+1} < P_n$, the pores whose radius is between r_n and $r_{n+1} < r_n$ will empty.

If the interval (P_n, P_{n+1}) is small, we can say that these pores have a mean radius given by:

$$\bar{r}_n = \frac{r_n + r_{n+1}}{2} \qquad \text{[A.20]}$$

If L_n denotes the total length of the pores of radius r_n and $\sum S_{n-1}$ the surface area of the already-empty pores, the total volume $\overset{n}{\underset{n+1}{\Delta}} V$ freed up in that interval is equal to the volume of liquid evaporated from the pores of radius r_n, plus the quantity of gas desorbed from the pores already emptied of liquid, so:

$$\overset{n}{\underset{n+1}{\Delta}} V = \pi\left(\bar{r}_n - t_{n+1}\right)^2 L_n + \left(t_n - t_{n+1}\right)\sum S_{n-1} \qquad \text{[A.21]}$$

However, if we let $V_{p(n-1)}$ represent the volume of already-empty pores with radius r_{n-1}, the surface of the already-empty pores is:

$$\sum S_{n-1} = \sum \frac{2V_{p(n-1)}}{\bar{r}_{n-1}} \qquad \text{[A.22]}$$

Furthermore, the total length of the pores with radius $\bar{r}_n$ is:

$$L_n = \frac{V_{p(n)}\bar{r}_n}{\pi \bar{r}_n^{\,2}} \qquad \text{[A.23]}$$

By feeding back the values given by expressions [A.22] and [A.23] into expression [A.21], we obtain:

$$V_{p(\bar{r}_n)} = \frac{\bar{r}_n^{\,2}}{\left(\bar{r}_n - t_{n+1}\right)^2}\left[\overset{n}{\underset{n+1}{\Delta}} V - \left(t_n - t_{n+1}\right)\sum \frac{2V_{p(n-1)}}{\left(\bar{r}_{n+1}\right)}\right] \qquad \text{[A.24]}$$

Based on the isotherm, then, it is possible to plot the curve $V_{p(\bar{r}_n)}$ as a function of $\bar{r}_n$ (with $\bar{r}_n$ being calculated using Kelvin's equation [A.6] and t_n by one of the methods indicated in sections A.2.2.1 , A.2.2.2 or A.2.2.3).

A.2.3.2. *Approximate method for very nearly vertical adsorption curves*

When the ascending part of the curve is vertical or very nearly vertical (Figure A.3), it can be assumed that for a given relative pressure, the total fixed amount is given by the desorption curve, whilst the adsorption curve shows the adsorbed quantity. This gives us our method.

The pores with radius $\overline{r}_n$ have the volume:

$$V_{p(\overline{r}_n)} = \underset{n+1}{\overset{n}{\Delta}}(\Delta V) \qquad [A.25]$$

However, in view of Figure A.38, we have:

$$\underset{n+1}{\overset{n}{\Delta}}(\Delta V) = A_n B_n - A_{n+1} B_{n+1} \qquad [A.26]$$

On the basis of the curves which give $V_{p(\overline{r}_n)} = f(\overline{r}_n)$, we can plot the distribution given:

$$\frac{dV_p}{d\overline{r}_n} = f(\overline{r}_n)_{\overline{r}} \qquad [A.27]$$

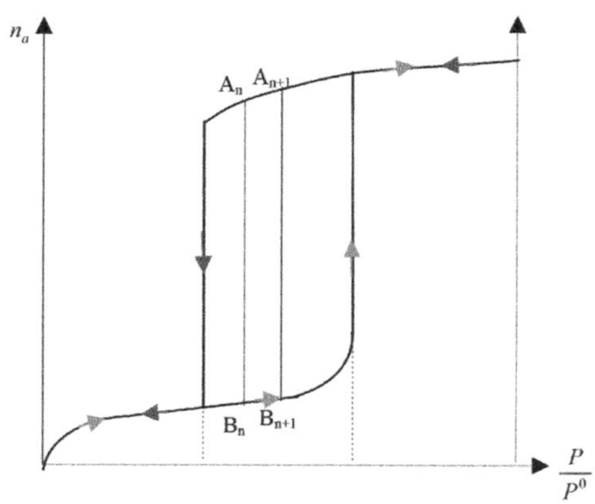

Figure A.2. *Hysteresis loop for nitrogen adsorption*

The adsorption method can be used to measure pore diameters of between 100 and 2500 nm. Above this range, the pressure ratio is too close to 1. Below this range, Kelvin's formula [A.6], which is a macroscopic formula, no longer applies, and hysteresis loops no longer occur.

NOTE.– We have seen another method to determine the pore size distribution, based on the infiltration of mercury (see section 3.2.2.2). In this case, we can measure volumes greater than 700 nm, so the two methods overlap in the 700–2000 nm range. Indeed, the results obtained match within 10%.

Bibliography

[ATK 90] ATKINS P.W., *Physical Chemistry*, Oxford University Press, Oxford, 1990.

[BEN 12] BEN-NAIM A., *Entropy and the Second Law*, World Scientific Publishing Company, New Jersey, 2012.

[BRU 40] BRUNAUER S., DEMING L.S., DEMING W.E. *et al.*, "On a theory of the van der Waals adsorption of gases", *Journal of the American Chemical Society*, vol. 62, pp. 1723–1732, 1940.

[DEF 51] DEFAY R., PRIGOGINE I., *Tension superficielle et adsorption*, Editions Desoer, Liège, 1951.

[DEM 73] DEMANGE H., GERMAIN G., NOTIN M., *Comprendre et appliquer la thermodynamique*, Masson, Paris, 1973.

[DES 10] DESRE P., HODAJ F., *Thermodynamique des matériaux, Equilibres de phase et métastabilité*, EDP Sciences, Paris, 2010.

[EMS 51] EMSCHWILLER G., *Chimie Physique*, PUF, Paris, 1951.

[FOW 49] FOWLER R., GUGGENHEIM E.A., *Statistical Thermodynamics*, Cambridge University Press, London, 1949.

[GER 59] GERMAIN J.E., *Catalyse hétérogène*, Dunod, Paris, 1959.

[GOK 96] GOKCEN N.A., REDDY R.G., *Thermodynamics*, Plenum Press, New York, 1996.

[HIL 56] HILL T.L., *Statistical Mechanics*, McGraw-Hill, New York, 1956.

[INF 06] INFELTA P., GRAETZEL M., *Thermodynamique, principes et applications*, BrownWalker Press, Boca Raton, 2006.

[KLO 08] KLOTZ I.M., ROSENBERG R.W., *Chemical Thermodynamics: Basic Concepts and Methods*, 7th ed., John Wiley and Sons, New York, 2008.

[LAL 06] LALAUZE R., *Physico-chimie des interfaces solide-gaz 1*, Hermès-Lavoisier, Paris, 2006.

[LIN 32] LINDAU G., "Über adsorption und kapillarkondensation", *Kolloid Zeitschr*, vol. 60, pp 253–263, 1932.

[MCQ 99] MCQUARRIEN D.A., SIMON J.D., *Molecular Thermodynamics*, University Science Book, Sausalito, 1999.

[MEZ 46] MEZGER E., "Loi de variation de la tension superficielle avec la temperature", *J. Phys. Radium*, vol. 7, no. 10, pp. 303–309, 1946.

[PRI 46] PRIGOGINE I., DEFAY R., *Thermodynamique chimique*, Dunod, Paris, 1946.

[REI 87] REID R.C., PRAUSNITZ J.M., POLING B.E., *Properties of Gas and Liquids*, McGraw-Hill, New York, 1987.

[SOU 68] SOUCHAY P., *Thermodynamique chimique*, Masson, Paris, 1968.

[SOU 90] SOUSTELLE M., *Modélisation macroscopique des transformations physico-chimiques*, Masson, Paris, 1990.

[SOU 10] SOUSTELLE M., *Handbook of Heterogeneous Kinetics*, ISTE, London and John Wiley and Sons, New York, 2010.

[SOU 11] SOUSTELLE M., *Cinétique chimique : éléments fondamentaux*, Hermès-Lavoisier, Paris, 2011.

[SOU 15a] SOUSTELLE M., *Phase Modeling Tools*, ISTE, London and John Wiley and Sons, New York, 2015.

[SOU 15b] SOUSTELLE M., *Modeling of Liquid Phases*, ISTE, London and John Wiley and Sons, New York, 2015.

[SOU 15c] SOUSTELLE M.., *Thermodynamic Modeling of Solid Phases*, ISTE, London and John Wiley and Sons, New York, 2015.

[SOU 15d] SOUSTELLE M., *Chemical Equilibria*, ISTE, London and John Wiley and Sons, New York, 2015.

[SOU 16a] SOUSTELLE M., *Phase Transformations*, ISTE, London and John Wiley and Sons, New York, 2016.

[SOU 16b] SOUSTELLE M., *Ionic and Electrochemical Equilibria*, ISTE, London and John Wiley and Sons, New York, 2016.

[VOL 25] VOLMER M., WEBER A., "Keimbildung in übersättigten Gebilden (Nucleation of supersaturated structures)", *Z. Phys. Chem.*, vol. 119, pp. 277–301, 1925.

Index

A, B, C

D, E, F

G, H, I

J, K, L

M, N, O

P, Q, R

S, T, V

W, Y, Z

Other titles from

ISTE

in

Chemical Engineering

2016

SOUSTELLE Michel
Chemical Thermodynamics Set
Volume 5 – Phase Transformations
Volume 6 – Ionic and Electrochemical Equilibria

2015

SOUSTELLE Michel
Chemical Thermodynamics Set
Volume 1 – Phase Modeling Tools
Volume 2 – Modeling of Liquid Phases
Volume 3 – Thermodynamic Modeling of Solid Phases
Volume 4 – Chemical Equilibria

2014

DAL PONT Jean-Pierre, AZZARO-PANTEL Catherine
New Approaches to the Process Industries: The Manufacturing Plant of the Future

HAMAIDE Thierry, DETERRE Rémi, FELLER Jean-François
Environmental Impact of Polymers

2012

DAL PONT Jean-Pierre
Process Engineering and Industrial Management

2011

SOUSTELLE Michel
An Introduction to Chemical Kinetics

2010

SOUSTELLE Michel
Handbook of Heterogenous Kinetics

Printed and bound by CPI Group (UK) Ltd, Croydon, CR0 4YY

13/09/2026

14961020-0002